D1784735

A Shepherd's Year

Dedication

To my brother Rod Hall, who was tragically drowned when he was 21.

A Shepherd's Year

Peter Hall with Vernon Wright

DAVID & CHARLES
NEWTON ABBOT   LONDON

Other books by Vernon Wright:

Stockman Country
David Lange: Prime Minister
South Africa

Hall, Peter
   A Shepherd's Year
   1. Shepherds — New Zealand — South Island
   I. Title II. Wright
   636.3 — 083 — 0924   SF375.5.N4
   ISBN 0-7153-9071-6

First published in Great Britain
by David & Charles 1987

©Copyright (text) Vernon Wright 1987
©Copyright (illustrations) Peter Hall 1987

All rights reserved. No part of this publication may be reproduced,
stored in a retrieval system, or transmitted, in any form or by any
means, electronic, mechanical, photocopying, recording or
otherwise, without the prior permission of David & Charles
Publishers plc.

Printed in Singapore
for David & Charles Publishers plc
Brunel House
Newton Abbot, Devon

Contents

Acknowledgements

The acknowledgements would be a small book of its own if I were to mention every person and the help they gave. Some have allowed me on their property or let me accompany them on a muster or they have just waited while I got in position with my camera, and yet others have done more by hosting me for days or giving me helicopter rides. However I would like to acknowledge the people that helped when I was just a photographer and not a musterer taking photos.

They are: Lorraine and Arthur Borrell, Guy Bellerby, Paddy Boyd, Jerry and Lesley Burdon, Robert and Linda Butson, John and Janet Cook, the Dagg family from Coronet Peak Station, Bruce and Merrin Douglas, Russel and Jennet Emmerson, Judy and Charlie Ensor, Bill and Mike Felton, my late brother Rod, Mandy and Mark Hasselman, John and Jill Hargest, John Kelland, Julie and Peter Lucas, Duncan and Carol MacKenzie, the Metherell family from Elfin Bay Station, Jim and Juliet Morris, Andrew Morris, Richard and Sue Murry, Colin Nimmo, Jenny and John Oswald, Wayne Parnham, Ted Phipps, Aaron and Maryann Radford, Liz and Bruce Scott, Andy Smith, the Snow family from Morven Hills Station, the Stevenson family from Upcot Station, Mike and Jane Thomas, and Ken Wigly from Glen Lyon Station.

My dogs, Peg, Guy, Tangy, Hitch and Bonny, have helped me above the normal degree of difficulty a high country dog is asked to perform, and I'm glad I can mention them here.

And lastly and most importantly, I want to thank my parents for always being there, my publishers for sticking with this shepherd for four and a half years, and Vernon Wright for writing the text.

Introduction

A *Shepherd's Year* is a book of my high country photographs and experiences as a shepherd. The book started in 1980 when I had a small collection of photographs which I realised wasn't just a record of where I'd mustered but was more an impression of my love for shepherding. And comparing them with photographs in South Island photographic books I noted that my mountains were alive with farming life and that the animals showed dignity and not just a desire to eat grass. I then decided I thought mine were better and I dared to wonder whether I should try and publish them.

To be sure there wasn't already a book such as I had in mind I looked through all the coffee table books I could. And I concluded that photographically the shepherd's high country was strangely neglected.

So for the next two years I carried a camera around while I was mustering and collected my high country on film, and then in the winter of 1982 I had to come down to earth and find out what publishers and other people thought.

The publishers' response to my then much bigger and better collection of photographs was very positive, but they were more cautious about the idea of publishing a whole book of them. The problem was that I hadn't thought about any sort of form or theme or even that such a book might need a text. They came to the rescue, though, and suggested the theme and form: a shepherd's year, and after I attempted to write a text they also suggested we find a writer. With that agreed to I went out again to take more photographs that would follow the "shepherd's year" idea.

About that time I heard of a *Listener* article on Glenaray Station, and being interested in anything to do with the high country I tracked down a copy. I was very impressed with it, it was up to date and full of interesting details. So I wrote to the journalist, Vernon Wright, and proposed that he join in "A Shepherd's Year", and to my surprise the man from the *Listener* said, "Okay".

I was surprised again after the text was finished. Right from January 1980 through to the year I got the contract, and through the next three years that I was working on it, I held my photos as great treasures and didn't think of the text in the same way. I saw it as a necessary part that I had agreed to have in the book but nothing more. But now I am very pleased to say that I think Vernon's text has brilliantly captured my personality and created an interesting and amusing text about a shepherd's year.

Peter Hall

SPRING

SPRING is shearing, tailing, green grass, lambs, more work and warmer temperatures. The great cliché of spring in farming country is of course the picture of an innocent gambolling lamb being fussed over by its mother. Sometimes I think a more appropriate symbol would be a sheep farmer concentrating on the weather forecast and hoping that that cold southerly isn't going to freeze the sheep now about to be shorn — and the shepherd standing there sympathising with the boss while wondering who is going to be in the gang this year. And of course there's more work to spring than shearing and tailing; there's also drenching, culling the sheep for wool quality and production, foot-rotting, droving, cattle-tupping, and careful management of the fresh grass.

Quite often spring means the start of a new round of work, perhaps starting work at a different place and forming new sets of relationships. During one spring shearing I ended up jumping the catching pen rails to get away from an angry shearer. My job was to bring the sheep through the shed to the catching pens for a fourteen-strong gang of blade shearers. Since it was a blade gang there was no machine noise to interrupt the talking, and I found

Spring muster on the way to Coronet Peak Saddle for shearing. I'm holding the front of the mob so the winter-weakened stragglers can catch up. Note how the sheep characteristically follow the contours of the land to conserve effort. Many generations of them, being mustered twice a year since the last century, have worn clear tracks into the hillside.

the chatter and general jostling fun. I became involved in this incident when the two-tooths (young, easy-shearing sheep) were coming up for shearing. The shearers had been working on the tussock-fed and shingle-backed wethers for three days and they were looking forward to working on the small and clean two-tooths. One shearer reckoned that the switch to two-tooths would increase his tally by more than ten an hour.

Then one of the gang, Paul, said to me quietly, "Pssst, come here . . . be a mate and cut our end out." What he was asking me to do was to give seven of the shearers easy sheep while the others innocently went ahead and finished shearing the tough ones.

Paul was the star shearer and a kind person, well liked by everyone, but what he was asking me to do was decidedly risky. And when Alan, the ganger at the other end of the board, saw me walking away from Paul and caught my eye, I knew very well what he must have been suspecting.

I had enough sense not to change one end before the other, but I couldn't resist implying I had. In a way I was getting my own back since Alan had been teasing me about what I did for girls away out here. I went back to the pens and shuffled the sheep around so that no time would be wasted filling a catching pen. Alan asked me directly, "Hey, are they on the two-tooths?" He paused, decided they were, and came after me. The sight of that red sweaty eighteen-stone hurtling at me was enough to call my bluff. "No!" I bellowed, as I flew away through the next pen.

A miserable Merino lamb in the spring. This one is about a month old and so is past the critical first few weeks when a cold snap can mean sudden death. Snowfalls right through to the end of November are quite commonplace — but it's not the snow that kills them: it's the accompanying cold winds.

It would be a pretty safe bet to say not many shepherds go to work by chair lift — though on the shearing muster at Coronet Peak they do. Here, shepherds from Coronet Peak are setting off from the top of the ski field after getting a lift up the hill.

IN August 1979 I went to work at a place called Cattle Flat, along the Matukituki River out back of Wanaka. Calling the place Cattle Flat was a bit like calling a giant Shorty, or a fat man Slim. There was some flat land around, but much of the mustering was done in the steep hills that rose up to five or six thousand feet above the Matukituki River flats. I was told I'd have to muster along goat tracks and ledges, and I was going to have to learn how to handle it.

I was to be one of a gang of four that shifted between seven stations. It was a semi-contract gang in that we got paid only for the days we worked. But the stations also guaranteed individual members work when they were not needed as a team.

At first I wasn't too worried by all the kidding about steep hill work. The cook at Glenrock Station used to refer to me as a mountain goat — I was used to steep hills, and lots of skiing had helped me get used to steepness. So I was pretty sure that, with caution on the slippery, tussock-covered faces, ridges, ledges and guts, it wouldn't be too bad.

But it is formidable country. The tales of dogs and sheep that have fallen off the bluffs are scarey. The teasing I was getting decided me to make out that I was very apprehensive about it all — and that attitude had some bearing on an experience in a valley called the Leaping Burn. The Leaping Burn was a mysterious place to me. You couldn't see into it from the farm, its opening was too high up.

The day had gone easily. Two of us had walked up and mustered back down the valley. For most of its way the going is gentle, but when you get to the end of the valley it drops off between 700 to 1000 metres down to the river, quite dramatically, almost straight. And I found myself following the sheep into the ledges and bluffs of that area. I'd actually called my dogs out of there because I thought it was too dangerous for them, and here I was wandering through the middle of it myself.

I stopped and sat down on a ledge in among the tussocks above and below me, realising that I was right in the middle of the area that my mates had been teasing me about, and that if I wasn't careful I could easily fall over. It was crazy: I'd called the dogs out for their safety and here I was trying to manoeuvre the sheep through on my own. And that of course increased the danger to the sheep, since their natural fear of me made it likely that they might fall off the precipice in their hurry to get away from me. At this place, the sheep had to travel along a ledge only six inches wide. The story is told of a dog on this particular ledge on a hill called Colin's Spur, trying to lift sheep up on to a safer path, then actually falling off the ledge and killing himself. And of course numerous sheep have fallen off it.

I decided to leave the sheep for another day. As the others agreed later, a live straggler is better than a dead sheep. Considering your own welfare isn't a bad idea either, once you've figured out where a bit of enthusiasm and pride have led you.

On another occasion in the same area the person mustering the hill above me, a guy called Kevin Short, came around a ridge and found himself among all these skiers racing past. The country is such that there is sun on one side and on the other — the cold side — there is a basin which is in fact the skifield known as Treble Cone. Kevin knew the skifield was coming up of course, but his dogs didn't, and their reaction to the skiers was a bit like a person's might be to Martians. They set up a barking and actually chased the skiers, although of course they couldn't catch them. But I had to admire Kevin. He went into the skifield car park and bought himself a pie for lunch. I thought that was rather good: on all other musters the top beat is lonely, the musterer is further out because the sheep are spread more sparsely, and he has only the skylarks to keep him company. It's not very often that he can pop in somewhere and buy a pie!

The intimate post-birth rituals of a high-country ewe with her twins. These lambs are only minutes old and still pretty unsteady on their feet. The mother very quickly licks them clean of the afterbirth and lingering moisture so their wool becomes an effective insulator at once.

High country sheep are not shepherded at lambing time; in fact spring is the one time of the year the ewes are left strictly alone. They are far too timid to be closely handled and the sight of a multicoloured shepherd emerging from the tussocks would certainly send them off at a gallop — or as fast as a newborn lamb can manage to run.

These pictures may seem to contradict my earlier statement that high country sheep aren't shepherded at lambing time. But the sheep here are Romneys, which originate from the Romney Marsh breed in Britain, and they have little in common with the high country merino from the arid mountains of Spain. They are farmed at the foot of these rugged, steeper hills, because the land is warmer and more fertile. Here, at Walter Peak Station, assistance is given to a ewe which has had birthing problems. At left, the shepherd is separating the ewes that have lambed, from the ones that haven't.

TAKING on the mustering job at Wanaka meant I had to get quite a lot of gear together. First up was a vehicle suitable for carrying my dogs, clothes and other gear. I had a ute with a specially designed top to carry the dogs. Some musterers carry their dogs in trailers. Either way dog boxes vary as much as the vehicles themselves; some are made of fibreglass and decorated with Footrot Flats cartoons; others are made of galvanised iron, or, like mine, of wood. Mine had an air scoop leading into the dog box. A further compartment, in front of that for the dogs, I used for stowing my other gear. The dogs got in through the tail gate. I never heard any trouble, fights, or niggles, if it was just my team in there. Occasionally they might snap at each other if they were tired. But in our travels we'd often come across a possessive dog. We'd put our dogs up in the farmer's truck and the next minute war would erupt: fangs, savage barking and lot of throwing themselves and each other, around.

My next item of travelling equipment was my walking stick, easily as important as a vehicle. The sticks are hand-picked from a manuka tree, selected for their straightness and lack of knots. When the wood is dry it's extremely strong, strong enough to spring you back from a heavy fall, and its smoothness and familiar curves make your hill stick much more than a mere piece of wood.

Boots are well looked after too. Mine were heavily oiled when they were new. I poured half a litre of neatsfoot oil in each boot. This ensures the uppers will have a long life. On the bottoms of their boots a lot of musterers still use leather soles and hobnails in preference to the much more easily maintained rubbers. I use leather soles because they are cooler and give a better grip on tussock country, though the grip depends on what steel is there. In the Raikaia, a heavily shingled area, boots are heavily nailed around their edges in order to stop wear. In Wanaka the slippery tussocks call for a biting grip. Some boots even have a two-centimetre length of saw-toothed steel jutting out to help with grip.

It seems all our gear is for travelling. We all carry a hill bag which will contain a parka, lunch, binoculars and perhaps a camera. I heard a delightful rule about carrying parkas: "Always carry one on a fine day and please yourself on a wet one."

I of course did carry a camera. Although there were many mornings on which I rose dreading the coming climb, saying to myself that I was doing more work than I needed to, and threatening to leave the camera and lenses behind. But invariably the reasoning would prevail that this day was going to be a smasher. And if I did get the shots I wanted then I'd pat myself on the back for persevering in taking the gear with me; and if I didn't get them, which was more often the case, I'd console myself with the thought: "At least you'll soon be lying on your bed with your bag off."

I always had to consider just what camera gear to take with me. From the time I rose at breakfast, made lunch and let the dogs go, I would be thinking, will I be taking scenic shots? If so, use the slow-grained film and the XGE body; or will there be close-quarters shots, such as loading into a vehicle — a helicopter, jet boat, or riding a horse? So the black and white camera body might come in handy. Carrying all this camera equipment, incidentally, earned me the nickname "Kodak". There was one lens I rarely went without: that was a 500mm mirror lens which I used with coarse-grained 400 ASA film. It was the snap-shooter for mustering where there is always such a distance between yourself and almost anything you see. It is such a powerful lens that I also used it as my binoculars.

Binoculars are not essential for mustering and many blokes on the hill don't carry them. However, I found my lens very handy for checking what appeared at a distance to be sheep — or for checking who was where and if all was going well. On one occasion I got quite a jolt from this. I was second to bottom in a team of five musterers stretched up a hill, and I was looking for the man below me. I was sure he was there, and through the lens I scanned the fern and scrub below to check. Into the viewfinder came a white backside. Such activities are of course perfectly natural out there in the hills. But sometimes binoculars pick up things you wish they hadn't.

Then there were always the smaller lenses. Will I need an extra wide angle to get those sheep in the foreground while the rest cross the bridge? Or the 100mm for portraits? Can I be bothered with the intimate involvement of a portrait? Maybe that heavy zoom will come in useful to pull the mountainous background in for a scenic shot? What is the day going to bring? . . . and speaking of that, I'd better hurry, because they're loading the dogs on to the truck.

Which brings me to the last two essentials for mustering — clothes and dogs. I wouldn't want to go without either, though I as good as went without my dogs once when I left my whistle at home. There I was with six dogs, all sensitive to my body language — for instance, the way you face is always the direction you want them to go — to the tone of my voice, and to my whistle, which could get them to do whatever I wanted — if only I could do it! How awkward I felt, knowing they could spin on a 10 cent coin at a mile — if only I could tell them! Instead, I was reduced to pointing and calling a particular dog's name, trying to kid them to go . . . and getting only bewildered looks in return.

Of course it doesn't matter what noise you choose to use as your dog language. Its effectiveness depends on how consistent the meaning is with the noise you make. Some aren't at all consistent. I heard one musterer describe a certain cocky's whistle as a "utility" whistle — meaning that he had only one. He'd whistle, so the story goes, but then his dog had to figure out whether it meant "left", "right", "out wide", or "go back to other sheep" or whatever — and he could only do it from the cocky's yelling at him. A sort of "you go and I'll tell you if you're warm".

I know of an interesting set of dog directions used by a dog trialist. He called his dog Trev, and he called Trev back with "You forgot your gumboots", he sent him to the left, as I remember with "What's the story?", to the right with "Fair go", and made him sit with "She'll be right." Can you imagine hearing all that at dog trials which for years have been hearing the same standard instructions "Come by", "Come in" and "Sit"?

ALL this is getting a long way from what a shepherd does during Spring in the high country. He rarely helps ewes with their lambing, because they are too timid for that, and there are just too many of them over too big an area. And when they are not disturbed they are naturally good lambers. The poor lambers don't survive, and that's all there is to it.

On those stations which shear prior to lambing, careful shepherding is needed. Basically this means putting the freshly shorn, pregnant ewes on the best grass, and if the weather cuts up rough, keeping them in shelter. It may be just under trees or in a hay barn that is empty after the winter's feeding out. We are constantly moving sheep — ewes to shelter, hoggets to the shearing and then to a block from where they can easily be reached for a drench if it's needed . . . or taking the wethers to the shearing shed, then maybe to rotate pasture after the ewes, to clean up the roughage the ewes didn't like, or we might put them out in the back country for the summer.

Everyone makes mistakes of course, and I'm no exception. Practically every New Zealand child grows up with the warning to close gates when crossing farmland. It's as ingrained (or it was!) as the rule to look right, look left then right again before crossing a road, and other injunctions of childhood. I well remember a gate being left open during a shearing muster at Coronet Peak Station, and the problems it caused.

We'd collected the sheep. They were in a big holding paddock

A Merino ewe with her newborn lamb. Merinos are still the backbone of the high-country sheepfarming economy, as they are the only breed to thrive on high-altitude tussocklands. They are mainly raised for their fine-quality fleece.

Walking out to the beats on a Glencoe shearing muster. It's September and still pretty cold, so the sheep tend to stay on the warm northern slopes. These shepherds are approaching from the south — the dark side, we call it. The point at which you first make contact with the sheep at the start of a day's mustering is important — you try and meet them in such a position that they naturally head off in the direction you want them to go.

Left: Cooking dinner in Round Hill Hut on West Wanaka Station. The fire also keeps you warm and dries out your wet gear. That cooking pot is made of cast iron so it keeps a very even temperature and is good for baking as well as stewing,

On a miserable wet day after the spring shearing, Jim Metherell is driving these sheep back out to their summer blocks. This is on the Greenstone River, where there are many river crossings and a lot of bush travel, so it takes two shepherds to control a mob which could easily be managed by one in more open country.

Well into spring and snow still lingers as Aaron Radford pauses on the top beat to view the grandeur of the Wakatipu landscape.

The Waters family, for whom I worked in 1980. I'm sure they won't mind my saying how very hard they worked on a farm they had down country so they could buy Glencoe, a high-country station. Here they are sowing a hay paddock on the station. Brian Waters is mixing the seed and "super" by hand in the drillbox, to ensure an even mixture is sowed. "Super" is short for superphosphate — fertiliser which helps restore the soil's fertility.

This rather roughly shorn, very hungry wether is snatching a mouthful of grass during a pause in the drive back from the homestead paddocks to its tussock block. Sheep are particularly hungry just after shearing because without their wool they feel the cold a lot more and anyway, they've been penned up overnight without much tucker. Down-country sheep often look a bit pink after shearing because they're closely shorn, but in the high country a bit more wool has to be left on them. This makes them look clean and lily-white for a few days, but they don't take long to get dirty again!

Once I tried to make a few dollars by getting my dogs to round up a few of the wild goats that abound in places among the high country. But they were too difficult to control, as the terrain was so rough and they really were completely wild — and to cap it all, it started snowing. So I gave up and settled for this picture of the mob being eyed by Tangy.

21

I had heard my father reminisce about the big mobs of wild deer they used to see while mustering, but now they are practically all gone; in one whole summer at Wanaka I only saw three of them. This is all that remains.

John Grant at the end of a day's hard work. (We'd got caught out in the rain.) He was in the Wanaka flying gang I worked with during the 1979-80 summer. The stick he's carrying is a "nibby" or hill stick, made from a seasoned piece of straight manuka stripped of its bark. The hill stick is used as a sort of third leg when you're working on steep country — and, occasionally, to whack a disobedient dog.

There probably are a few questions that should be answered here, like what's he doing? Well, we were tailing lambs and while we were drafting off some lambs from their mothers this ewe was caught. It had missed the pre-lamb shearing two months before so we gave it to Tiger to shear, but he just went and sat down on us and had a smoke!

A mature Merino wether (castrated ram). He shouldn't really have horns like this; they should have been debudded when he was a hogget. This is because although the horns look grand they are murderous if they catch you in the back of the leg when you're working in the yards. And they're difficult to shear around.

The yellow ear tag is a colour code to tell the animal's age. In this case, yellow indicates that he was born in 1975, and since this picture was taken in 1980, he was then five years old.

Wethers at Glencoe Station being driven back to their tussock blocks after shearing. Controlling
a mob of 3000 sheep like this can be difficult and the very simplest of mistakes can cause big
trouble. Just stopping the lead too quickly will cause the mob to concertina, and if the fences on
either side are not strong enough, the sheep will flatten them. Often, if you're at the back of a
big mob you may be three or four kilometres back from the head and you can't even see them.

which stretched from the Coronet Peak skifields road down to the bottom terraces. The shearing had started and we were to go out in the morning, drive up the road and collect the rest of the sheep to keep the job going. But when we got up there we found the gate open.

There really couldn't be anything worse. There was nothing to stop the sheep from going back to the country we'd just mustered. We walked down to the gate, trying to ignore the shock by talking about why and how the gate had been left open, how many sheep were likely to have got away, and so on. But we all knew the sheep would be as far out as they could get on country they knew. Once they had discovered the gate was open, they'd soon have let the rest of the sheep know they were on the move, and once out they would just keep going until they were back in the country each liked, to their own little patches.

For me the worst was yet to come. There were only two of us who might have left the gate open. It was an old gate and we could have argued that it should have been a better one. But the fact was, we'd tied it up and it had come undone. There'd been a fierce wind the night before too, but nevertheless all these excuses came down to the fact that we — or one of us — had tied the gate up and now it was undone. It was a Taranaki gate (i.e., made of wire and fence battens) and the top had come loose and collapsed. The sheep just walked over it. We dreaded having to muster that whole country again. It really wasn't much fun — it was a two-day job, no Sunday walk.

Shutting gates becomes an automatic reaction and it took me some time to come to terms with the fact that it could have been me who was responsible. I wasn't deliberately trying to forget, and I made a mental note of my reluctance to accept involvement in that kind of mistake, just in case I was tempted to forget it.

DROVING is one of the most enjoyable jobs, and a lot of it happens near the end of spring when the paddock grass is getting down. The wethers are the first to be moved, as the ewes are kept close in for tailing and, later on, weaning. They are moved along roads, then tracks, and then up spurs.

This is the time to encourage a young dog. Young dogs don't need to be stopped as there isn't much harm they can do. They are unlikely to stop the front of the flock as that's miles ahead, so they can bounce up on some sheep and send them scuttling along the line. It's important that the dog enjoys it. And it is awfully funny watching him learn. He'll bound up to some sheep, give a sudden woof, the sheep will get a start and fly off, and the pup, startled himself, will jump back with surprise. Or another time he will find he's cut into the mob and can't get out. The sheep part and make way for him and he will shuffle out looking self-conscious — a bit goofy.

It's pleasant, too, working your older dogs without the necessity for the usual strict control. They end up doing spontaneous little things. They detect sheep creeping off the track so they quietly position themselves where the sheep will walk into them, and they take great delight as the sheep jump to get away from them. Or they simply patrol the side of the flock, enjoying every second of not being told what to do. All we have to do is follow. There are no tactics or co-ordination with others as there are in mustering. The creeks have to be watched so smothers don't occur, but otherwise it's just a matter of following.

The other job near the end of spring is tailing. It's a chore in strict contrast to droving as it calls for strategy and strict control, and it is tense. The strategy is to muster the sheep and keep them coming off the hill slowly and steadily so that the potentially wild lambs will stay with their mothers. The control needed is to keep your dogs calm and thus the sheep quieter. That's how it is meant to work. But often groups of lambs do get mis-mothered and

Despite its forlorn expression, this dog is actually serving a useful purpose by being placed at the back of a temporary tailing pen, where it keeps the sheep from pushing against the weak fence. This also provides an opportunity for the dog to take a well-earned rest.

This is Tangy. Apart from being a hero to me she is a typical heading dog. That's the type of dog that brings sheep back under the power of its eye. Without any instructions from me, she will run silently out over many kilometres and carefully approach the sheep, win their confidence and bring them back successfully — all on her own.

She's quite independent and likes controlling the sheep in exactly her own way — if I give her too many commands she'll ignore them. It's not that she doesn't understand; she simply isn't prepared to be pushed beyond her own limitations. Of course, this independence of hers is an advantage — whereas most other dogs need constant attention, I can leave her all to her own devices for a couple of hours at a time.

when the mob is brought together the tempo picks up, the pressure increases, and these lambs want to break free. And if they do break free the whole system can break down: one person turns his attention to the lambs and his out-of-control dog, and the main mob finds a gap and goes through it.

The tailing itself is probably the nearest thing there is out there in the hills to a factory assembly line in town, where each person has a specific, repetitive chore. Someone picks the lamb up, another earmarks it with the station's registered brand, others drench it, cut its tail off, and then if it's male, they castrate it.

SHEPHERDS are keen dog trialists, and sometimes they almost get to compete while going about their normal chores with dogs. On one occasion it just so happened that I'd had to make a difficult run with a dog in front of a workmate by name of Paul Emerson, nicknamed Tiger. My run wasn't particularly technical but the dog did do it well. He ran up and headed off the sheep that were moving along a fence line nearly a kilometre away. Then there were some sheep further away still, and I had a special command that told the dog to go out again. I called it "recast", some people just call it "go back" — and get more sheep. He went out neatly, got some others and brought them down the fence. Everything was done well and in front of a stockman who was eight years my senior, a man with a good reputation for his ability at competing in dog trials.

Half an hour later I was doing my beat, which went around above Tiger, looking on to some paddocks down a steep drop of about three hundred metres. Tiger was standing there moving sheep from the hill. The bluffs in that area were like a jigsaw puzzle, with the sheep scattered among them. Tiger sent his dog up to get the sheep, and as they were quite a long way away he just sat down. That meant the dog couldn't get any indication of direction from Tiger's body; all he was getting were whistles.

With my dogs, as with a lot of people's, it's very much an arm signal which gives an indication of which way you want a dog to go, in addition to whistles. So when Tiger sat down it was quite obvious that his dog knew when to go left or right, and also to what degree he should go; a big left, or whatever. That in itself was quite impressive. It also meant that Tiger had confidence, he could just sit down and let his dogs go about their work. It was a show, and it was well done. He sent his dog up along little tracks that were difficult to see unless you were actually on the spot. The dog was going in and out of these little bluffs covered with fern and rubbish. The sheep seemed to be reluctant to emerge so he had to force them out the way he wanted them to go. I think they wanted to lead back out of there and back behind me.

Normally in a situation like that you would leave the sheep to their own devices to get out. There's no need to read the country yourself; you just let the sheep lead and you follow them. But they wanted to go in the opposite direction, which was probably the way they'd come into it. Tiger, though, tried to force them back the other way, and this meant some intricate work at a distance of nearly two kilometres, moving the dog around, keeping everything in tune. The dog was wanting to co-operate and was using its own quietness to influence the sheep, and he was taking commands from Tiger as to which way they needed to be pushed. It was a really impressive show, and there was a kind of competition there for the space of an hour, which was enjoyable for both sides. We both sort of said, "good run."

IN late spring you sometimes get to handle big mobs of sheep, taking them out to the back country. On one such occasion, at Glenrock Station, there were 3000 in the mob, all wethers. We were driving them along a road with fences on either side. The blocks of fenced paddocks would end eventually and the wethers would be in open country with no fences to keep them in. So one

New and old systems for handling the lambs while they are being tailed. Above:
Lambs are slid into the trough where they are immobilised while being inoculated,
drenched, earmarked, castrated and tailed. On the right they are getting the same
treatment, but because they are hand-held the job is more labour intensive.

Left: A typical tailing yard, built on the block in just half an hour.
There are two good reasons for doing the tailing on the block. One is that it's
difficult moving uneducated lambs — by that I mean ones that aren't used to
being worked by shepherds and dogs. Also, more importantly, the lambs need to
be left alone and on clean ground after docking to reduce the risk of infection.

of the men ran a dog out to stop the mob just near where the fences were to end. This would cause the sheep to group up and we could position ourselves and have a better chance of controlling them. In the meantime, as the front of the mob was stopped the back grouped up — and up, and up — and eventually they flattened the fence. The pressure accumulates so quickly. You can't point to just a few sheep in the mob doing the pushing, they're all doing it, and the force becomes irresistible. What we did wrong was not reading the situation. We should have pulled off the dog stationed in front more quickly so the pressure could be relieved. The sheep didn't damage themselves, as it happened, but they could easily have done so.

Sometimes dogs can get themselves into a pickle, too. I recall once during a muster when I'd come down off a steep bluff in order to save time and get in behind some sheep, all my dogs except one followed me. It was pretty risky and the dogs had to be careful; if they fell, it would be for about five or six metres. But this dog got only halfway down. To help him out would have been encouraging him to be dependent on me, and if that was taken to the extreme he'd be yelling for help at every steep bit we came to. So I left him overnight to think about it. He stayed there until the following day, and I decided that he was just too old to change his ways. I did go up to get him, but my attitude was that if he had been younger I would have left him there for two or three days until he came to realise that he had to take the responsibility for looking after himself.

Once dogs realise that they've got to look after themselves they can make decisions: they see a bluff and *they* make the decision as to how they should handle it. They can go around it. When I'm mustering I just haven't time to worry about the dogs' dilemmas, so they learn early on that I'm not going to wait; they've got to find a way to keep up.

It works well, usually. The dog works to its own capacity, it knows the rules and respects them. I've seen people take pity on the dogs and go back. I think that's false compassion in that the dog then seems to get into bluffs more, and still gets stuck.

Different dogs have different abilities. I had one that just wouldn't have anything to do with bluffs. I accepted that he wasn't up to it, and so he always went round them rather than through them.

On one occasion, at Braemar Station in the Mackenzie country, I'd been doing a beat sidling along the side of a hill and I got very sore feet. I had had it by the end of the day. To add to that, near the end of the day I came across some woollies (sheep that missed the last shearing), that needed to be brought in, and that required more energy. The following day we had what turned out to be a long walk across flat ground for something like sixteen kilometres. It seemed to drag on and on and on, and I was getting a bit sick of it. It was a hot day and there were really no sheep to trace and I had to force myself to cover the country.

It was one of those times in which you go over and over little things in your mind and you have to try really hard to persuade yourself to move. I was bored. Halfway through the day in this state I saw what appeared to be a log in a pond. I thought it looked a bit strange — a black and white log! I was working a dog at the time and when I'd finished I walked along a bit and then noticed the "log" was moving. I got a bit closer, then realised that it was in fact a Canada goose with her young. The goose had its neck right out, laid flat, eyeing me quietly. It was paddling, moving away from me, and its young were doing the same thing. It was delightful to see, to watch my influence on it and to see it protecting its young, for otherwise it could easily have flown away. A small incident perhaps, but it made me feel better.

When sheep are jammed together in the yards, the occasional daring lamb like this one will make the most of the opportunity to escape. They will walk or run across the platform of sheep, showing initiative and daring, then clear the fence and make their bid for freedom. John Moore has managed on this occasion to recapture an escapee, but the outcome can be quite different if two or more simultaneously head for the hills!

A woollie being handsheared. This one's missed at least four musters and so hasn't been shorn for two years or more. The wool fibres, though very long, will be of uneven thickness along their length because the fibres grow thinner in winter owing to the shorter food supply. So the wool's of poorer quality.

Inside the woolshed at Coronet Peak Station. Within two weeks all of the 10,000 sheep on the place will be shorn here.

Right: Cecil Peak is one of a select few South Island stations that do not have road access. Here the wool clip is being unloaded in Queenstown Bay, with the steamer Earnslaw on the left and the Remarkables in the background. Each bale weighs about 150 kg and the wool goes to a big sale in Dunedin.

SUMMER

WHEN summer rolls around almost the first chore is hay — cutting, carting, and stacking it. Most of the hard work has been taken out of this now with the big balers and forklifts. Three stations up the Rakaia, Glenrock, Glenariffe and Double Hill, combined their trucks and men and we competed at hay carting. It was not just a matter of who worked the hardest as both the truckloads and the loads in the barns had to be bound so they didn't collapse when the last bale was put on. The edges had to be kept tight and each layer put slightly inside the other so that it leaned inward. And in the middle of the stack adjoining seams were never repeated. The best time was when the stacking was finished. We'd all sit around and enjoy a beer.

In the lead-up to Christmas there were a few special community events. Up the Rakaia it is the annual barbecue, and at Glenorchy it is horse races. I have competed in two horse races, one at Glenorchy and the other at Mt Nicholas, which is more of a gymkhana. Both are very informal. And both I remember because of blunders. The horses are amateurs when it comes to racing, as are the riders with their long stirrups, lack of bright colours and hard hats. That year there were about thirty-five horses, and a lot

At the junction of the Dobson and Hopkins Rivers, on the first day of the annual Glen Lyon calf drive. The drive takes two weeks and covers 180 kilometres, from the cattle yards near the shores of Lake Ohau, passing through the townships of Tekapo, Burkes Pass, Fairlie and Pleasant Point, and finishing in the Temuka saleyards where the calves are sold for fattening.

more riders than that. So the horses were kept busy. I was busy photographing but I thought okay, I'll go for a ride. I didn't have a horse, but it didn't matter.

The last race involved fifteen two-person relay teams with one horse each. The race consisted of the two members of each team racing for two hundred metres, one on foot, the other on horseback, then changing places and running back the two hundred metres. Of course, the horses outrace the runners and so their riders have dismounted and are waiting to swap over by the time the runners get there. You're all lined up, fifteen people on the ground and fifteen on horseback, and then you're away. It's a miracle no one gets trodden on, because the horses are nervous and we're all milling around each other trying to get a good start in behind the horses. It's pretty chaotic.

One runner was only twenty paces behind the horses, which was pretty impressive. And I was running about three chain behind him, feeling rather pleased with my efforts. But there was trouble ahead, as I hadn't ridden a horse for over a year. I'm reasonably confident around horses though, and I wasn't worried: I'd sussed the horse out and he was calm, he wouldn't worry about me jumping on too fast.

But I did jump on too fast, and we were away at a canter in about two steps. I didn't bother about stirrups, I only had two hundred metres to go and I didn't think I needed them. And about halfway back I fell off! I could see it coming a mile away. My legs felt jelly-like from the running and just couldn't get any grip. I

35

Haymaking is a big job at Godley Peaks Station. There are 22,000 bales to make, and this take about 80 hectares of irrigated paddocks. In winter, the hay is fed to 10,000 sheep and 150 cows. Without the summer hay crop it would be impossible to winter-over so many stock because of the snow which often covers the ground, and the altitude and frosts.

A recent innovation on many big stations is making huge round bales which are handled mechanically and save a great deal of back-breaking work.

The Glenorchy "local gallop" is on and John, Barry and Bruce are well out in front. Race day at Glenorchy is held on the first Saturday of the New Year, and events include the walk, trot and gallop race, the trotting race, the relay, the sprint, and the ladies' gallop. The races are held mainly for locals, though friends from further afield do get invited. Glenorchy is a small town with a pub, a post office and a garage, situated at the head of Lake Wakatipu.

bounced like a rubber ball and I fell off. Determined to save a bit of pride I held on to the reins and each time I fell off I'd jump back on. Of course, it cost us the race, although we still got a placing.

The other horse race incident occurred the previous year at Mt Nicholas Station on the south side of Lake Wakatipu. A pony club from Te Anau was invited and all their members came in, right from tiny tots to adults. Again, I was in a relay, with two other Peters — Peter Goodall and Peter Chilwell. This was more a traditional relay, in that each of the three team members had to ride a third of the distance, in turn, on the same horse. I recall a bit of conversation before the race: I asked Peter Chilwell if the girth was tight, and he said, yes, it was. Peter Chilwell was to ride first, me second, and Peter Goodall finishing it off. But it didn't get to that.

Almost as soon as they started I could see that something was amiss. Peter was leaning too far forward. He was still in the group, but something was wrong. He was now lying along the horse's neck, they were halfway down the straight and three hundred paces off. Other riders were glancing over at him. Then he let go the reins and wrapped his arms around the horse's neck. He was yelling "stop, stop" — and then he slipped around underneath the horse's neck and was clinging to it with his arms and legs. Other riders managed to stop the horse just as it reached us. Peter let go and looked around, a bit embarrassed but otherwise surprisingly unconcerned. The girth hadn't been done up tightly enough and the whole saddle had swung around underneath the horse. We lost the race of course, but we had a good laugh about it afterwards.

ANOTHER leisure activity is salmon fishing. When I first tried fly fishing I thought it was bit tame, but the first time I hooked a salmon I was shocked at the size of it. The rod bent over, the reel screamed as the line tore out, and I had to use some strength to hold it. After a few losses I learned how necessary playing the fish is in the process of actually landing them. And then I really enjoyed it. Being right next to the river at Glenrock Station meant we could duck down after work for a few hours, or even before breakfast. And if the river was too clear or too dirty and the fish weren't taking, we didn't waste any time trying. We had it made — a millionaire's sport at our back doorstep.

Glenrock is a good hour's drive from any farming community, so it is quite isolated. On one occasion I well remember the station was visited by a pretty American girl. There were three of us working there as shepherds, all about the same age, and one morning the Ensors, the runholders, decided to take us and Katie salmon fishing. So we drove out in the truck across the river bed and up and along the Rakaia. The three of us were excited at the prospect and anxious to show off to Katie. I actually jumped off the truck before it had stopped, such was my eagerness, and I ran to the river thinking, I'll show her, I'll cast my line first and get a fish. And damn it, I did! First cast of the season. I couldn't believe it. It happened so fast that the others still hadn't got down from the truck. I was in hysterics.

Then I lost the fish. I was using a small bait caster borrowed from Charlie Ensor, the boss, and I didn't know how to work it. One of the others, Hugo Sandal, caught a fish on his third cast. I think our visitor was impressed. We were quite surprised ourselves. And then I got another strike! By this time Charlie was on the other side of a backwash and I called him to come over and show me how the bait caster worked. Charlie came roaring across the backwash to help me land the fish — he could see my rod bucking and bending — but it was very deep. Charlie disappeared, thigh waders and all, right up to his neck. And then he actually started drifting away. He must have been close to the main current. I was still fighting my fish, and thought it all a bit of

Far left: *The huntaway is New Zealand's own sheepdog. This one, Tod, is a typical example: large and placid with black-and-white and tan markings and floppy ears (though here they are standing erect in the wind). He has a powerful, deep bark and a natural instinct for keeping sheep on the move.*
At fifteen months of age, Tod has just become a mature working dog, but some dogs mature earlier, depending on the skill of the master. Some, I'd say, are never properly educated!

Left: *Moving sheep from the upper Shotover River down to the Wakatipu Basin takes about two days and covers some pretty dramatic scenery. This remarkable stretch of the road is called Pinchers Bluff and was constructed by gold miners in the 1860s. It's a sheer drop of forty metres to the river but doesn't seem to worry the sheep a bit.*

Right: *There can't be many farmers using flying foxes to get to their property, but this is how the Felton family, of the aptly-named Mount Difficulty Station, cross the turbulent waters of the Kawarau River. Here we are just coming back across from making a check on the stock and the amount of feed they're getting, at the top end of the block.*
Mike Felton is coming across last with two of his dogs and the one behind him is pretty nervous about this mode of transportation, while the one between his knees is really enjoying having such close contact with the boss. Many dogs are scared of travelling on a flying fox, and I can't say I'm too keen on it myself!

a joke. I wasn't aware of any danger. It all happened so fast. There I was with a fish on the line, trying desperately to get it, watching my boss bobbing up and down, and still excited by the girl's presence. Charlie found his feet and pulled himself out, but he was not amused. And the fish got away. I didn't catch anything myself in the end, although we took home two fish. I think our visitor was impressed with the fish, if not the organisation.

THERE are other sports too. A cricket match with the local stock firm, and rugby in the winter. Then there are huge valleys and wandering tops to explore, each one as new as the one explored the weekend before. At Glenrock there's a broken wall of bluffs up from the farm, with guts separating knobs, and behind that a tussocky plateau I've only been told about. There is a valley called Petticoat Lane, in behind Double Hill, that looks like a gigantic shingle pit except for its curves and distinguishing rock outcrops. There, a kea's call echoes so much you have no idea whether the bird is in front of you or somewhere behind.

I once went on a horse trek up the Wilberforce River, which joins the Rakaia, and saw Mt Algidus Station, where my father had spent enjoyable days as a musterer. And then on up to Browning Pass. We left the horses there and followed an old pack route up a steep face to where it passed through a bluff, and we were awed at how dangerous it must have been driving cattle through there. The cattle went mainly to feed goldminers on the West Coast, so I was told. Then we sidled away from the bluff and climbed a steeper alpine mat of moss and lichen and many other species of plant to view a lake surrounded by lilies — Lake Browning.

Hunting was also fun, especially planning your stalk and handling the rifle. The game I shot mostly was hares and rabbits around the flats, and chamois on the hills. Deer were too few to worry about. Mike Paul and I went on our motorbikes down the Rakaia Road seven miles from the Glenrock homestead to try and get a chamois. We turned off the road at Redcliffs and rode up a little-used track to Cookies Flat. Cookies is completely covered in tussock, unlike the hills with their rock outcrops, matagouri bushes and sharp line soft shingle slides. It is a treasured place to me. Few people visit it and the plateau, some eight hundred metres up, would have remained unchanged for many years. It was a pleasant change from the fences and pine trees of the Canterbury Plains.

We were heading for the back of Mt Hutt, an area dominated by its huge shingle face fourteen hundred metres high. From the top of this face we could see the Pacific Ocean, as well as a lot of the Canterbury Plains. Musterers are good at combing the hills with their eyes — they have to be of course — picking up the slightest contrasts of colour or shape, and movement. Even if the movement isn't actually seen, a change in position will register, especially if the object has been consciously noted. I spotted them first, but I had quite a job convincing Mike that the tan specks we could see were chamois. But they were, and there was no other way to get to them but straight up that huge shingle face. Up we went, hoping their eyes weren't too sharp today.

It was tremendous climbing with the anticipation when and where were we going to take a shot? Would I shoot well? Where would the ones not hit run to? My rifle was a new model Parker Hale with a floating barrel and stock. It had three extra qualities: the breech had a lip for better hold, the pistol grip was lengthened, and the butt had a rubber pad. The prize of the rifle was the scope, an American Weaver with steel barrel, the best glass, and four-power magnification.

I had the weapon. I only had to use it well. That was, if the chamois didn't see us first. We climbed for half an hour and there they were, four hundred paces away. We moved a little closer, shifted a few rocks in the way and then lay flat with our rifles pointing up. Mike counted down and we fired together and struck

Cows and their calves pause among the sun-drenched tussocks on their way down to the Greenstone Valley from the Mararoa Block. They will be held in the Elfin Bay Station home paddocks on the shores of Lake Wakatipu, where the calves, which are about 6 months old, will be weaned. These are good examples of Hereford cattle, which are a long-established beef breed.

Loading and carting hay is backbreaking, sticky, repetitive work. Here at Godley Peaks, with Lake Tekapo in the background, the merciless summer sun begins to take its toll; limbs ache, sweat trickles and dust and chaff get in everything.

When a lamb peels off from a mob, it's called a lamb break. This is most likely to happen when they've become mismothered — that is, separated from their mothers in the mob which, as you can imagine, makes them pretty upset. Lamb breaks are dreaded by shepherds because they can start a chain reaction, with the entire mob rapidly getting out of control. The way this bloke is threatening the lamb with a stick reflects the tension of droving a large mob of ewes with lambs.

Here I'm following one of three cuts out of a mob of 400 cattle on their way down to Elfin Bay. It's the third day of the muster and we're nearly there. Now that the breeding season's finished, the bulls aren't at all interested in the cows and are quite content just to meander along at the back.

Droving cattle through river crossing after river crossing is usually no hassle, but when they have calves at foot it's quite a different matter. Then, normally quite biddable old cows will really object to being pushed by the dogs. The cows stamp their feet and try to butt the dogs in an attempt to protect their calves, and sometimes the dogs will bite them on the nose.

Taking this picture was risky: the bull had just been taken from his mob and he was a bit upset about it. Nor was he too impressed with me for ever-so-casually bending my knees to get on eye level with him.

Matt Parsons with a renegade lamb that tried to escape the weaning muster on Killermont Station. This one broke away from the mob and was stopped by a quick heading dog, though Matt would have had to do a fair bit of running too!

We've been in the saddle for an hour and half already, but it's still early morning as Lake McKellar comes into view. We are going to muster cattle and drive them back down the Greenstone River, which drains the lake.

Here a fence line is being planned on the Black Spurs to keep sheep from going up into the high mountain basins in late autumn. John Hargest is measuring out the distance with a length of chain during a passing shower.

The end of another day, and Jim Metherell, on his horse Cracker, leads three footsore dogs through the last river crossing. Horses are indispensable for cattle mustering when there are calves around, because they don't upset the cattle the way dogs do. And, of course, the horse carries you as well.

The Rats Nest is typical of the huts where shepherds shelter when working in the high country. It's a spartan but serviceable one-room dwelling with bunks and an open fireplace for cooking and heating. Guy Bellerby is coming up from the Greenstone River after a quick wash at the end of a day's calf marking.

Bulls are quite excitable at mating time. In the weeks before they go with the cows their usually huge fat bodies become lean masses of muscle. They butt heads and have head-to-head pushing competitions. Then they roam around the cows in what is for them the very serious task of mating.

This is Bloke, a five-year-old huntaway/beardie belonging to Jim Metherell. With working dogs you need constantly to be thinking ahead to how hard the work will be later in the day, and working them accordingly. There are many factors to consider like whether they'll just be following sheep, or fighting angry cattle, and how hot it'll be or whether the surface is rough going.

If you manage your dogs well they'll always be fresh and willing, and if you don't they'll become disobedient and slack on the job.

*Early morning mist lingers on the flats and terraces of the Mararoa River as Jim Metherell
heads out for the day's cattle muster.*

the rocks below the chamois. They were on their feet and away in the time it took me to slip the rifle bolt back then forward again. The second shot landed at their feet. the third one I didn't see, but by then they were out of range. They climbed steadily away to the left and quickly became tiny tan spots once again.

I have since found that hunting with a camera is more lasting fun because your shots don't frighten the animals away. And if you are a good stalker and the wind is in your favour you can get amazingly close. I have been within a stone's throw of two deer.

I enjoy travelling through the hills but I also enjoy breaks away from it, like going to the Christchurch Show for the week. This is a great show because people from all corners of the high country go to it, and often it will be your best chance to meet old friends. It is also a chance to catch up on things like who's paying what wages to whom, where there might be a job coming up, how a particular farm is developing under its new owner, or that so-and-so has got married. It's a good time to make contact with friends and get some knowledge of places you may not have seen.

SOME events in a shepherd's life are caused by his, or someone else's, foot being planted firmly in the mouth. I recall a friend called Tony who had just returned from a holiday with another musterer and their girlfriends. They'd been on a two-week tour around the South Island, and Tony had put on a bit of weight. At the time we were about to start the weaning muster at Cattle Flat, where the climbs are quite severe because of their suddenness. We were having our breakfast when Tony remarked casually that he might be slowed down by his extra weight. It was just a little bit of modesty, however. He knew as well as I did — and most of the others — that he had real determination, a real aggressiveness if it came to it, to overcome any unfitness or extra weight he might be carrying. So I took what he said with a grain of salt. One of the other musterers said to Tony at breakfast that

we'd go slow for him. I thought then that he should be careful what he said.

Anyway, later on during the climb we had stopped to take off our jerseys. Ahead of us lay an exceptionally steep climb which would take about half an hour before we began a gentle descent into a valley, and from there we would go separately and muster out and down the steep front face. To my horror someone suggested to Tony that he might as well take the lead since he would be the slowest. He couldn't have said a worse thing. Tony wound up, and before long none of us had any spare breath for talking. I can still see Tony running those words over in his mind, 500 metres up the hill, and he just went. I was leaning on my hill pole in an attempt to push myself up. I no longer noticed the dogs playing, or the sun rising, or the breathing of the others in the early morning air. I worked only on following, because I wasn't going to be left in the awkward silence of regret. I was carrying the usual amount of camera gear, too, as I stuck in behind him. The pace split us up. I managed to stay with Tony, but it took an awful effort. I'm not sure if the others couldn't keep up or they didn't bother to try and it was just Tony and I who were bull-headed enough to keep up the pace, driven by pride or whatever it is that drives you. By the time we stopped at the top there was a gap of several minutes before the others caught us up. There was no talk then about Tony's condition, and nothing was said about how anyone was feeling. Tony had made his point, and the rest of the journey was conducted at a civilised pace.

The weaning muster is probably the most trying muster of the year. In winter there may be snow to wade through, and in spring the sheep may be weak from winter, and even in the autumn when the sheep should be strong and you need plenty of fast dog power, it's not as bad as coping with the midsummer sun. The saying "An hour in the morning is worth three at midday" expresses this well.

You are so surprisingly powerless. It took seven of us and heaven knows how many dogs more than three hours to move two thousand ewes and their lambs just a stone's throw down the hill to a paddock at Castle Flat during one weaning.

They wouldn't respond. I'd act as berserk as possible, flapping my arms, scraping my hill stick wildly on the dirt or beating it on the fern, and all the time encouraging my dogs to bark by whistling and shouting "speak up here, here here speak up". It had the effect of moving twenty or thirty ewes and lambs a few paces. If you compare this to sending thousands running just at the sight of you, you'll understand why I felt so powerless.

However, there was a connection between their response and the amount of noise they'd already been subjected to. I learned to handle sheep as little as possible and as quietly as possible, so the wave of an arm or walking towards a sheep kilometres away would send it off.

I once disobeyed this rule because I was mad with the boss. It was a Sunday. I was in love, and she was to ring at seven that night. In the morning I told the boss I had an important call coming and I wanted to be in by then. I rather suspected my request would be ignored, but it wasn't as if we were out at camp or mustering all day; it only meant arranging my jobs so I'd be at least close to the house then.

In the morning we drafted a few sheep and troughed a few rams for footrot. Since I hadn't started taking the mob out before lunch I assumed I'd leave after lunch, but it was two o'clock before I got away.

I lined the sheep up on the track, put a huntaway on both sides and worked them back and forth, keeping the sheep trotting.

Some people think sheep must be pretty fit from all that roaming about in the hills. But when you consider that the musterer and dogs may go up and down and all over the block all in one day, they are really the fit ones.

Perhaps I might make it after all. There weren't any weak sheep among them; in fact they must have been fit since they went at a fast walk for most of the afternoon. But with only a kilometre or two to go and no track for me to squeeze them on, they stopped. There was no puffing or cramp, they simply weren't used to this handling and they weren't going to put up with it any more.

I hadn't struck anything like this before. I had six healthy, effective dogs; I could pull sheep down from a mountain, cross them over a river and put them up a scrubby ridge on the other side, with just a few whistles. Yet here I had sheep right in front of me and they wouldn't come out of a circle. I'd edge a few off only to have them run back again. It was hopeless, and since we were going out that way again the next day, and as there were no sheep on the block, I decided to leave them. They could be moved later. Besides, it was after seven and I wasn't so interested in being a dedicated musterer that I was prepared to ignore that phone call. In fact I was far from dedicated, or so my boss told me when he picked me up fifteen minutes later.

The sheep I'd left were mustered on to the next block the following day, and in doing so I got a quiet reprimand from one of my dogs. Two of us scooped the sheep off the one hundred hectare block and put them over a steep little creek just inside the next block. It was there that I noticed my heading dog was missing. Panic set in. The block the sheep were on had been poisoned for rabbits. It was safe for sheep, but if a dog ate as much as a leg of poisoned rabbit, it would die.

I left the other guy and rushed back whistling and calling as if my life was at risk, as well it might have been, in a sense, if I had had to muster those hills without that particular heading dog. I called and called for about five minutes. Then, apparently quite unconcerned about my state, the dog brought in half a dozen sheep from a blind gut just above me. I felt terribly small — we'd missed them, but the dog hadn't.

51

It's lonely working the top beat on a muster. The next shepherd may be a kilometre away and you've got to think ahead as you move along, to ensure you keep in line with the others and cover the country. This means a lot of walking up and down, and using binoculars to search the land ahead of you to ensure you don't waste time covering areas where there aren't any sheep.

Above: A shepherd driving a small mob on down towards the next bloke so he can carry on unencumbered along the top.

Right: On steep hill country like this your hill stick is a real godsend; we call it the third leg. Of course, sheep are completely used to the steepness and are also at an advantage in having four legs.

These sheep aren't jumping over a wire or anything — they're leaping high as a kind of release of tension, as they realise they're finally getting through. They will have been waiting perhaps half an hour while Mark Hasselman counts them through the gateway in single file.
It's a good sign when they jump like this, because it means they're fit and well.

Sometimes a station's best grazing is isolated by road. Here Coronet Peak sheep are being driven along the many kilometres of narrow road which separate the warm slopes of the upper Shotover from the homestead, where they will be dipped and crutched.
Dipping is done to kill external parasites, and crutching is the term for clipping the long wool away from both ends of the animal, so at one end it doesn't dirty its wool and at the other end it isn't blindfolded by it.

Droving scenes.
Left: A lot of the work is just hanging behind the mob for hour after hour like this.
Above: The end of a hard day, and thirsty sheep reach the shores of Lake Wanaka.
Right: River crossings are tricky. If the sheep go across in too thick a bunch they literally act like a dam; the waters back up above them and sweep them off their feet. You have to keep close control over the sheep and string them out into a narrow line. These sheep are being driven across the Dobson River from Glen Lyon Station to Huxley Gorge.

THE next event of the summer is calf marking. Cattle work is very different from sheep work. We herd them, drive them and handle them as we do sheep, but I find that the subtler whistle tones that give emphasis or say what speed to push or pull, are just not used on cattle work. It may be that way because cattle give great threatening bellows at the dogs, exciting them and making subtle control very difficult. Either way, cattle work looks like a brawl compared to the careful persuasion of sheep work.

This is particularly so at calf marking. The cows have fed their sleek calves since spring and are very attached to them. It's magic to see the effect a couple of barks have on tussock cows. In the Mararoa Valley I did this and the place came alive with bellowing mothers. Out of scrubby guts, from high terraces, out of the bush and from the large snow tussocks on the river bed they came, each concerned to find the calf she had left with a foster cow, and determined to protect it. If I hadn't known they were charging down to their calves and not me, I would have bolted.

Once they have found their calves the air settles. But the first conflict about moving, where dog and cow meet head on, causes another mountainous bellowing session, with many cows running toward the conflict to lend a hand. Of course, the protest is against the dogs more than having to move, and once they are moving and the dogs have quietened down then the cows settle down too.

Cows will charge people if they are upset enough. A head-shaking four hundred kilogram cow once charged my eight-year-old sister. Several of my family were moving sixty or seventy cows up a paddock. Somehow, one angry cow left the dogs and came charging down on my sister. It happened too fast for anyone to help. She froze and screamed; the cow ran to within a pace of her then stopped, turned and ran back to the mob.

Being in the yards with irritated or anxious cows can be frightening, too, given their size. I found that sympathy rather than anger works with cows. After all they are only nervous — of the yard's high rails and metal gates swinging before and after them, nervous of being packed in and nervous of the huge din they're making. It's quite different from normal life on the hill.

It's easier to calm your own nerves by rushing around shoving cows here and cutting them off there and generally working hard. But you'll get the job done easier and probably quicker if you slow yourself down and set a quieter mood.

In one set of yards the boss' son, who was about twenty-four, watched me do this then said I was lazy. Sure, I was only spending a fraction of the energy he was, yet I was still drafting the cows as quickly as he was firing them up to me. And he looked so ridiculous. A cow would pass behind him and let the air out of its lungs; suddenly he'd swing around, whack it, and get kicked in return.

At Elfin Bay Station in February 1981, I worked with Bob Metherell, the runholder, and Guy Bellerby, marking around four hundred calves without as much as getting kicked once.

It was a very enjoyable time. The yards had been built in the bush using big old trees as posts. So we were roofed by trees. The back of the yard extended deep into the bush, so it was covered with moss, everything was very green. The yards themselves, with the movement, the steam rising from the animals, and the calling of the cows and calves, were a lively place. I enjoyed watching the hierarchy exercised by the bulls. At the top of the hierarchy an old bull would wander carelessly around smelling the cows, while at the bottom end a young bull would jog around keeping away from all the others.

The very start of the Glen Lyon calf drive to Temuka, as the 1800 calves are released from the yards. The calves have never been driven on their own before and are difficult to handle, especially while they're fresh. The idea is to drive them hard for a while to tire them out a bit, then once they're out in the riverbed they're more submissive and the renegades become easier to push around.

This young heading dog could be saying, ''Why walk when no-one's riding the horse? Besides, I'm quite partial to some Kris Kristofferson music!'' Along with three shepherds, their dogs and 700 calves, the young dog is heading for Tekapo during the Glen Lyon cattle drive.

Two of the shepherds shelter from a nor'west wind at the edge of Lake Pukaki. This is one of the rest stops during the drive, to let the cattle move at a more leisurely pace for a while, feeding as they went.

At night in the hut there, called the Rats Nest, Bob, Guy and I were woken by Bob's eighteen-year-old son working his dogs in his sleep. There were some plum adjectives included and Bob had to yell at him to shut up.

It's a wonder I haven't started shouting at my dogs in my sleep myself, given the time I've spent wanting my dogs to work better. When I started mustering, the experienced men sounded really articulate and the dogs looked so responsive. How did they get so good? It is only now, after having trained dogs myself, that I realise teaching them isn't a magical trick but a laborious job.

I teach the basic commands before trying to direct them on sheep. The commands go, sit, left and right are all done by first whistling then physically moving them forward for the "go" or pushing their backsides down for "sit". This is done over and over until they anticipate you moving them. My dog work doesn't compare with the precision required by a dog trialist who directs his dog a few feet to the left or right at a flat gallop, but it earns me a living.

And then there's your own training. The older I get the more I look for fault in what I'm doing and try to change it when I find it. Funnily enough the dogs benefit from this more than anything. My own difficulties with change have made me sympathetic to their faults.

I can clearly remember the first formal chipping I got. I had run a dog in haste to collect some sheep that had been missed by myself and the guy on the beat above me. They were already behind the line of the muster, so if I botched it they wouldn't bolt down on to someone else's beat. What I did wrong was to ask too much of my dog. I didn't walk back, but sent him up on his own, and when the sheep spotted him coming they rushed further back and I was in no position to stop them. I was told all this quietly in front of four others as we walked home after the muster. The word "position" didn't have a lot of meaning then, but I sure remembered it and worked at finding a meaning for it. Now I see "get in position" as meaning that you should run a dog from where he has a good chance of succeeding in what you want him to do.

I was helped another time at the Methven dog trials in March 1977. At any trial you will find four courses and three classes for the dogs. Two of the courses are to test huntaways and two to test heading dogs. The huntaways push three sheep up a face, usually a hundred to a hundred and fifty paces up. They do this on courses called straight hunt and zigzag. There are only two markers on the straight hunt and they are right at the top; the zigzag has two other sets marking the zigzag.

With the heading events there is a long pull and a short head and yard. The long pull is where a shepherd stands in a ring marked with stones, about ten paces in radius, and brings the sheep down the hill to him; it finishes when the shepherd, the dog and the sheep are all perfectly stationary for at least a second.

The short head and yard starts the same way, but at the ring stage you have to drive them down a one hundred pace course through a set of hurdles twelve feet apart, and thence to a pen. At the pen you open the gate and branch out from its end to wing the sheep in. You're not allowed to let the gate go. The dog must bring the sheep to the pen and push them in with your help as a wing on the gate.

The three different classes, open, intermediate, and maiden, are judged by the same person, and the points system links district trials to the South Island and National championships. Winning an open is five points and qualifies for the national championships. A second is four points and needs one more point for qualification, and so on down to fourth.

I was a long way away from thinking about qualification in March 1977, so much so that two respected local men surreptitiously helped me compete. It was my turn on the straight hunt. My dog was a fiery, aggressive type whose bark would send the most sluggish, mud-puddling Romney shooing along. The trappings of the contest were all new to us — the cars, the old bar, the judge's box that looked like a giant long-drop except that it had a door and a window, and the V-shaped pen that led to the three-sheep release pen.

I was in front of the V-pen with my back to the judge and a few onlookers. And I had just had my sheep liberated for me. The direction I gave the dog was to stand, which seems a little silly since he was doing that anyway, but it also meant to him, "hold it, you're on." A barrel of barks made me jump, and sent the sheep up the hill. And just as I was about to start some pushing a voice behind me said, "Hold him there." Then another voice said, "Keep him well off here at the start." They were telling me what to do. Nothing that I was about to do was as subtle or as patient as what I eventually did do. My clandestine advisers told me to hold the dog further back, they held him even when the sheep were drifting off line. They seemed to know if the sheep would go uphill and back on line without being pushed. I was very proud of the performance, even without winning a place. The men had given me the feel of their experience.

In that trial I was like a transmitter between my dog and those men, and I have actually used walkie-talkies to do a very similar thing during a muster. It happened with Charlie Ensor on two separate occasions, when each of us lost sight of the dog we were running and we instructed each other through walkie-talkies as to what we should tell the dogs. I told Charlie that he'd missed sheep and if he ran his dog, Bonnie, I could tell him where to send her. Off went Bonnie back up the ridge they had just come down and into a basin that was out of Charlie's sight. Bonnie wanted to swing through the basin, but I kept her going up until she spotted the sheep. It was pretty simple, she hooked them, and it worked perfectly, apart from having to stop her pushing them too fast. The sense of power was amazing: there I was, two kilometres away, thinking to myself, "if they slow down they'll sidle out at a better level", and saying "Stop her, Charlie", and seeing it work just as I'd planned.

It was only a couple of days later that Charlie responded in kind. I'd been winging sheep up a face and meant to bring them down, but they rounded a ridge before my huntaway got a head on them, and that's when the walkie-talkie in my hill bag piped up and said "Keep him going." Rather reluctantly I whistled "away." What mess was this going to lead to? "Stop him, he's put in a good head ... that's it, keep steadying him, out to the right, great, he's doing a good job." What? A good job? He's never done that in his life before, surely he's not doing it on the only time I trust him out of my sight? But I enjoyed the experience — and the sheep *did* come back on to the ridge and down to me.

Lambs which have just been drafted from the ewes for the last time, at about 4½ months of age. They're pretty nervous and almost impossible to control. These are replacement stock, which will be kept in well-fenced paddocks until they become disciplined to being handled by the dogs. This requires intensive handling for about a week. Then they'll be put on to the best, most succulent grass. Nutrition is particularly important at this stage: food has to be good quality and plentiful to replace the mother's milk.

An older wether — this one would be six years or so. Up to five years you can tell their age by counting the teeth. In the first year they only have milk teeth, and every year after that they get two more strong, prominent, permanent teeth until they have ten such teeth in their sixth year. At these different ages they're correspondingly known as two-tooths, four-tooths, and so on, until they finally become called full-mouths.

Most shepherds take part in dog trials, and enjoy the mixture of competition and socialising. Here Arthur Eason, a well-known trialist, carefully approaches the pen he is to yard the sheep in on completion of a competition called the short head and yard, at the 1984 Wanaka dog trials.

The competition starts on the other side of the two large gates. From there he sends his dog along a curved path for a few hundred metres to sort of hook around three sheep. Then the dog takes control and brings the sheep back between the gates, and on down to the pen. It's all done working against the clock, too, so there's a lot of pressure on. The stick at right is a fancy, regulation-length mustering stick.

Dipping using a shower. Sheep are dipped to kill parasites that destroy the wool or suck blood. It also protects them against blowfly strike. Many modern sheep dips are so concentrated you just pour a bit on each animal's back, which makes the job easier and is less stressful for the sheep.

These lambs are not needed for wool production or breeding, so they have been sold and are being loaded onto a truck. Loading is strictly controlled so the animals aren't overcrowded, and they are carefully counted into each deck on the truck. This also provides a double-check on the tally.

AUTUMN

THE mustering job to beat all others is the autumn muster. It is the climax of the year, an end to summer signalled by bringing the sheep off the tops before the snowfalls make it impossible to reach them.

For me it is a great adventure and challenge as well as a worry — and even a threat. The change of people, sheep and scenery makes it an adventure, and the worry and the threat is slipping up somewhere and missing sheep, whether it is through poor use of a dog or not linking up with the others.

It is the time of the year in which to see quite a few different stations. When I worked up the Rakaia I was permanently employed by the Ensors at Glenrock for two and a half years, and they worked in with Double Hill and Glenariffe Stations for the autumn, so I got to see three places there. In Wanaka I was employed on "The Run" and so mustered for seven different places. They arranged when and where we worked and the station mustering paid our daily wage. In the Wakatipu where I was casually employed, I spend most of my autumns on Mt Creighton and Coronet Peak Stations, with a few days at Elfin Bay, Temple Peak, Walter Peak, Ben Lomond and Wyuna.

On a day like this, when the air is light and fresh and you've been lifted in by helicopter, you feel on top of the world. Here we are on the Glenorchy Mountains at the head of Lake Wakatipu. It's autumn muster; time to bring the sheep down before the snows of winter. Underneath all that cloud there's the town of Glenorchy, the lake, four sheep stations and half a dozen farms. The high peak in the background at right is Mt Earnslaw.

Autumn means a lot of walking. To write about that seems unnecessary, it must be almost the simplest thing we do. But really, I have had vastly different experiences walking. Some of them I would rather forget, some have been funny and yet others are cherished memories. They have been different mostly because of the different people I have had to follow. Some have wanted to race, and I fall into the category of, while not wanting the hard work of racing, not liking to be left behind either. Still others have no idea of what pacing is. I got quite a shock when I walked in a trail of six people led by a head shepherd who really made walking easy. At the start of an eight hundred metre climb I was towey. It was only because he was the head shepherd and fifteen years my senior that I didn't rush ahead. I was feeling quite restless, his pace was so monotonous. About halfway up and having settled down I noticed that our walking was queer, or at least I felt queer. My legs and whole body were fixed at a pace that seemed to perpetuate itself. There was no desire to speed-up or slow-down or stop. I felt as if I could walk forever.

I've heard joggers describe this same feeling. I have no problem in climbing the highest hill, whatever state of unfitness I might be in, as long as I can use that same slow monotonous pace.

However, when you are one of the employees you don't often get to set the pace. I had a real mix-up in this department during one of my first days mustering on Coronet Peak Station in 1981. Brian Dagg and myself were climbing a hill. This can be quite a competitive activity. You don't actually set out to race each other,

The autumn muster can mean climbing to the top of mountains like these then battling to control very fit sheep and in the process cutting your dogs' pads to bits on the sharp rocks. Although this ground looks pretty barren, at about 1600 metres, there are actually little scattered basins about where there's some grass. This is about as high as you ever have to go. Any higher and you'd need oxygen!

Bill Dagg still insists on taking his turn riding on the back of the truck with the dogs, even though he's been on Coronet Peak Station for twenty-five years and has two sons running the place as well as a number of much younger employees on the job. He's using a horse cover to keep the rain off.

because that wastes energy, but sometimes it comes close to racing.

On this occasion I was following Brian. It's my habit to follow fairly close behind. I've never thought about why, I just feel comfortable that way. I can switch off and all I do is put a foot exactly where one of his has just been. He thought I was pushing him. He went a bit faster. So I went a bit faster, just because I was trying to follow close behind him. And it got a bit faster, and faster. At first I didn't think much about it, but then my chest started pumping and my legs were beginning to complain. We were still talking meanwhile; the atmosphere appeared quite casual, and this went on for three-quarters of an hour up nearly a thousand metres of steady climbing. When we got to the top we were both really puffing. He showed me where my mustering beat was and we went about our day, both knowing we were tired but neither of us mentioning it.

It wasn't until two or three days later, when I was making some statement over a beer about racing being for poor musterers and fools, that Brian nudged me and said, "What about Advance Peak?" He said then that he had been pretty stuffed: "You were pushing the pace a bit, weren't you?" And of course I said no, I thought *he* was pushing the pace, and I was just trying to keep up. We had a good laugh about it.

I BECAME confident at working sheep. I enjoyed the challenge of working with them and most of the time my control was good. I haven't always fared well with cattle, however.

During a spell at Glencoe Station I worked in a team of four people. I was probably the most able person in the team, and that gave me confidence. The more difficult the work, the better I liked it, because I enjoyed the challenge. In addition I had at that time a team of six good dogs, and that always helps.

On this occasion we were collecting cattle in the Cardrona Valley. The boss, Brian Waters, was going up a side creek and I was to wait on a river fan for about forty minutes until the cattle came out. I was to turn them up a valley and they would be taken home from there for calf marking. Well, I sat down and went to sleep. I didn't think it was such a bad thing, and I had hours to wait. I knew that I would hear the cattle as soon as they came, and of course the dogs would be barking. But it all went wrong. I thought I would hear them a lot earlier than I actually did, but they were past me and already heading back when I woke up and panicked.

I had enough sense not to try and run dogs. I thought I'd run out towards them, that I'd be able to get there and turn them back. But I couldn't. They ran faster than me. I kept going for about a mile or so, in among the leaders, trying to pass them. By that time I was sending dogs out but they were being ignored in the general shamozzle. I still fancied my chances, that if I could get ahead and give them time to see me they would stop. So I crossed a little creek we were running parallel to, jumped a fence and did a loop. I got about one hundred metres ahead of them because they slowed down once the dogs and I weren't so close. I threatened them with a stick and gave them time to consider. But they considered they could get past and they kept on coming. At this stage I knew I'd failed and it was very embarrassing.

It was made worse by the fact that the boss was very meticulous in his directions of what you should do, and often he seemed almost over-instructive. Over a period I tend to heed my own counsel: I've done the work before. I can handle it, no trouble. And there I was having done just that, not taken a lot of notice of his saying to make sure the cattle didn't break back. The boss was not an easy-going man, either; he would be disappointed. And all this was reflected in my frantic efforts to stop them.

The end result of it all was that about half a dozen of the cattle ran back — cows that were looking for their calves. Early in the

Shepherds often affectionately call the areas they muster "the hills" when, as shown here, the word "mountains" would plainly be more appropriate.

Left: Straggle muster on the upper Rakaia. If the tallies aren't what they should be, shepherds are often sent out to round up the strays. Here Jim Morris and I have found a few stragglers just after the first snowfalls in May. That's the main divide of the Southern Alps in the background.

Above: Lucky musterers have just stepped out of a warm Jet Ranger helicopter and are about to muster the Twenty-five Mile Creek in the Wakatipu area. Without the chopper it'd be a 2½-hour walk before even starting the day's work. There actually are sheep grazing on that incredibly rugged face in the background.

Right: Brian Dagg watches his dog working sheep, from the vantage point of a knob at the head of the Shotover River. Even from more than kilometre he still has control over his dogs.

morning some cows will leave their calves with a keeper cow. Maybe half a dozen calves will stay with one cow, while the mothers will go off to graze. These cows that were charging me down were likely to have been the ones that went off grazing and were now frantic to find their calves. When the herd was brought together the mothers probably didn't take the time to look over all the calves that were there. They just took off. Usually a cow will go to retrieve a calf from a keeper cow as soon as it hears a dog barking, but on this occasion they obviously got muddled.

Brian decided that they would be all right because their calves would be returned the next day, after they'd been marked. They would find each other and not a lot of harm would have been done. All the same, it wasn't what he'd planned and he wasn't very pleased.

I WROTE a short story once, an autumn story called The Adventure of Burdon's Cup. It started with a man falling over a bluff and ended with me winning a dog trial while on crutches. It occurred when we were mustering the Mt Burke wether country between Lakes Wanaka and Hawea.

There were six of us doing the long valley country, mustering off approximately six ridges, usually taking a ridge each. On this occasion though, I was doing one ridge with the boss' son, Tim Burdon. He was just out of Christ College and hadn't done much steep country work or mustering. He had three good dogs that his father had bought him, but little things showed him to be not too confident on the hill. The ridges were really quite steep with odd little areas of bluffs that you had to find your way through. When you haven't been to a particular place before you're confronted with the sheer problem of knowing where to go. From experience I've learned to look for sheep tracks, even if the track just shows as some dead tussock leaves brushed to one side, and from such signs I judge whether an imminent bluff is passable.

I was working above Tim when I heard a cry, and then a loud thump. There was a chill up my spine. How could I get to him? I knew there was a bluff between him and the creek below. I ran to see where he was, and sure enough there was the bluff. I couldn't get down, I couldn't even see him. I yelled out, but there was no reply, which didn't make me very happy. I went back up the ridge I had just come down, so I could see better. This was about two or three minutes back and I was running, so I could see the other ridges that ran out to the left — a distance of about a mile. I radioed his father and said Tim had fallen, I didn't know what condition he was in; stand by. No doubt my voice conveyed the concern I felt. I then rushed down a little rock chute, taking a gamble that it was the shortest way down. I got to the end of it and was confronted by a four-metre bank. There was a tree halfway down it so I thought I would grab hold of that. In doing so I slipped and whacked my knee against a rock, but I kept on down an incline towards the creek. It was steep, a difficult place to run down.

But halfway down the hill there he was, half sitting up and trying to gather himself. My immediate reaction was, great, he's all right. At one stage I'd had no way of knowing if he was even alive. For me to have been able clearly to hear the thump meant it must have been a considerable fall. So I was amazed that he was moving around, and I guessed that there was probably not too much wrong with him after all. He said, "Where's my dog?" I more or less said, "Bugger your dog." He said, "He fell down after me." I knew immediately which dog it was. It was a big, heavy dog that wasn't suited to the light and nimble work needed in and out of bluffs. I thought then it was rather funny that the dog was silly enough to follow his boss down.

Tim was dazed, his hands were quite knocked about, and he was off work for two weeks. I thought he had been rather lucky, because with only a little bit more rolling after his first landing he

This old scheelite miner's hut has a great view of the Glenorchy areas, where I'm taking part in
the autumn muster. I'm carrying a light daypack: cameras, parka, lunch, and that's about it.
Some shepherds don't even bother carrying lunch, but get by on just two huge
meals a day. I developed the odd habit of saving the crusts out of my sandwiches,
just in case I got delayed after dark. Quite often that's come in handy, just keeping me
going for that little bit longer!

To be a shepherd, you need good lungs — for breathing hard when the hills are steep going, and for yelling when your dog is too far away to hear you (or if it doesn't want to!). I'm not sure which category Bruce Watson's dog belongs in here . . .

Important basic instructions are given at the start of a day's mustering on the Mid Run in the Wanaka area. We were new to this particular property and got advice on where to hurry or slow down, and how to keep the sheep away from the favourite escape routes they'd try to use to out-manoeuvre us.

Mustering off the sheep we've rounded up, at the end of a day on Elfin Bay Station. We took them down to a holding yard for safe keeping while we mustered more blocks for a couple more days before droving them all on down to lower country.

Outnumbered a hundred to one, the dog (at left) still has complete control because it has been taught to treat the mob as though it were one animal. As an inexperienced pup, the dog would have bowled into the middle, widely scattering them, then probably singled out one sheep and chased it to the bottom. Rigorous training and preventing bad habits from developing is the key to rearing an effective sheepdog.

Once sheep become used to a swing bridge, like this one across Moke Creek in the Wakatipu area, they have no trouble crossing. In fact, you have to keep the gates closed on either side or they'll wander freely back and forth when they aren't meant to.

It's not often sheep get themselves into such a ridiculous situation as this! These ones have missed the previous autumn muster and wintered in a mountain valley, then escaped the following spring muster. Now they are perched on that rocky knob in a last attempt to stay for yet another season. That attempt was unsuccessful — we managed to drive them down.

Sheep on the skyline of a leading ridge on Mt Aurum, Coronet Peak Station. At night sheep camp on any flat ground they can find — and in country like this, that often means moving on to the crest of a ridge, where they really need the protection of their wool against the elements.

Smoko in Passburn Hut, on Elfin Bay Station, and Bob Metherell is pouring the tea. We had bread and jam to go with it. The hut's built of beech poles and sheathed with corrugated iron carried in by packhorse. The floor is dirt, with old sacks serving for a carpet.

Sheep stream out of a holding paddock at the start of the autumn journey down-country. To get this picture I had to make some very undignified sheep noises to try and stop one or two of them while the others rushed by. All the baa-ing earned me a few raised eyebrows from the other shepherds!

This beautiful stretch of water is Lake Luna, on Mt Creighton Station, and we're driving a mob of wethers past on the way in from their summer block.

Bob Metherell looking for stragglers in the Sly Burn, a tributary of the Greenstone River. You need to spend a fair bit of time systematically looking for movement. Often you will take a second look after a few minutes at something you weren't quite sure about, to see if it's moved. Binoculars are useful for checking suspicious white rocks and "vegetable sheep" — a kind of plant which is light in colour and grows as a mound about the size and shape of a sheep. Vegetable sheep are incredibly deceptive.

This beautiful crayon sketch was drawn on the back of a hut door by Don Ward, a British goldsmith who was working as the cook with a mustering gang. He did this in the course of an evening after finishing his day's work.

Autumn muster at Mt Creighton Station. One of the things shepherding has going for it is that every day you work in the kind of country that tourists pay the top dollar just for a glimpse of.

would have gone down a grade much steeper, and once started he wouldn't stop. Then, when he got to the top of this ridge some seven or eight hundred metres from where the accident happened we met Cotty, Tim's father. He just told me to be sure that I stopped the sheep from bolting down the valley we'd just come up.

I was by now limping quite heavily as a result of hitting my knee. The fall had happened at about eight o'clock in the morning and it was 10:30 or 11 by the time we met up with Cotty. I carried on mustering and by the end of the day I was more concerned about myself than I was about Tim or how lucky he had been. I had a damned sore leg. I think Cotty probably was very worried, but he just didn't want to talk about it too much.

At the pub the following weekend I was telling what happened. I then had bandages on — not plaster, but the effect was the same, I couldn't bend my leg. Of course I looked a lot more injured than Tim did. No one, it seemed, was prepared to acknowledge just how close he had come to losing his life. They said, "Oh you should have let him go." They were very off-hand.

Anyway, the dog trials then came around. I had six dogs and they were all working, so I thought I would enter them all. We achieved a complete in the long pull. That's where you take the sheep from a distance, in this case about four hundred metres up the hill, and the dogs have to run out in a straight consistent line, and then in a consistent curve to get in behind the sheep, and then bring the sheep back to where you're standing in the centre of a circle. I put two dogs in that event and got a complete with one and about halfway down with the other, which gave me some points.

With the yard trial — not as far to head as the other, but the same rules, a consistent run out and then back to you — the dog goes out eighty metres. Then you and the dog manoeuvre the sheep through pegs set out in a line, and then through gates until you get to the end. They are then put in a pen made up of four

gates. I didn't get a complete, but both my heading dogs' runs got points, and I felt rather good doing it, hobbling away on my crutches.

And then I entered three huntaways in two events. I got five completes in all, and found out a couple of days later that I'd won the local mustering cup for aggregated points. The cup was the G.L. Burdon Trophy (G.L. was Cotty's father and Tim's grandfather), and I felt I particularly deserved to win it.

THE high country is frequently an arena where only the fittest survive. That applies to animals as much as to the people who look after them; perhaps more so. I was once mustering a small block on Mt Creighton Station with a nine-year-old girl who went to school in Invercargill. She was up on holiday with her mother and she accompanied me on a mustering beat. We came across a cast sheep that had obviously been there for a long time. I put it on its feet and waited for its muscles to revive, holding it upright so that its balance would come back. In the meantime the muster was carrying on, and the girl asked me if I was going to come back and get the sheep or stay with it and make sure it was okay. Normally — that is, on a big muster — I wouldn't have had time. You soon realise on the hill that there is only a limited amount of time in which to do a certain amount of work, and you have to make quick decisions. But there I was, working to try and save this sheep, and all the time thinking how little idea my companion probably had of what it means to be a sick sheep way out here, and what your chances of survival are, as opposed to a sheep's chances if it's further down on the flat, where the farmer can go and pick the cast sheep up, and where keas don't dig holes in their backs. In the extremes of extensive farming, only the fit survive. This sheep was very weak, and more than likely would have normally been killed for its own sake. Usually, if it's badly cast, you try and right it, but in the long run you might as well

The first time I took my dogs up in a helicopter, I thought they'd think it was some alien craft and be frightened out of their wits. But (to my disappointment!) they were only slightly frightened and I could see their fear would soon vanish. And sure enough, in no time my heading dog was looking out the window for sheep, while the others carried on socialising just as though they were on the back of the truck.

This beautiful old stone wall is part of the Skippers Road. The sheep climbing up the bank at the back are licking salt off the rocks to get their trace elements, which may have been deficient in the grass they've been feeding on.

This Temple Peak wether is trying to get across without wetting his feet. But there wasn't quite enough room for his back feet as well as the front, so he fell right in and got thoroughly soaked.

Some shepherds carry a shearing blade with them so they can shear any woollies they come across, rather than bring them in. Woollies are sheep that have missed shearing at least once, and they may be carrying a lot of excess weight which weakens them. Here, Len Lake is bringing in a woollie's fleece at the end of a straggle muster on Branches Station.

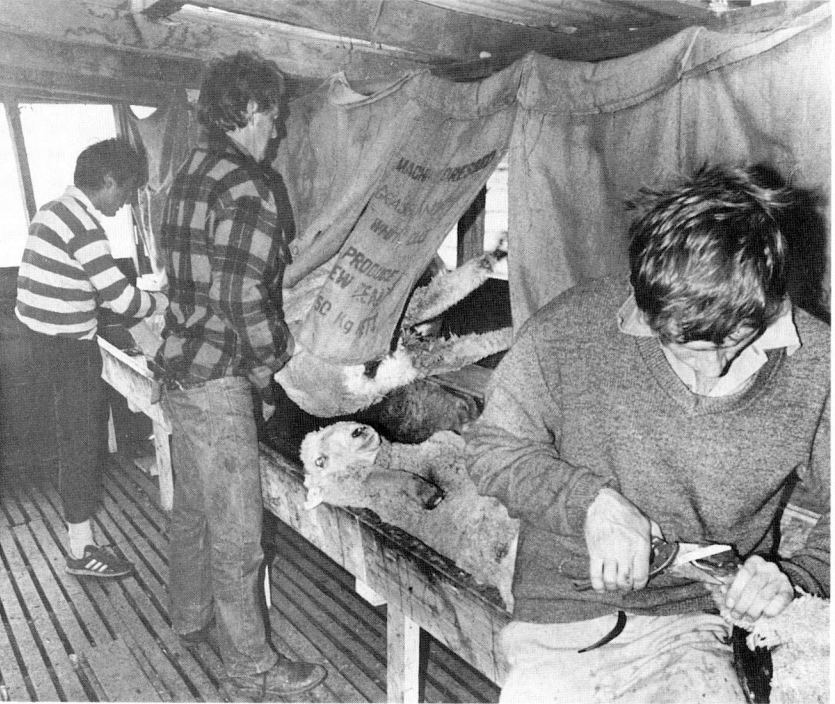

This is foot-rotting. Foot-rot is a serious infection that particularly knocks fine-wool sheep by rotting their hooves and stopping them moving around and eating. The foot-rotting job is a tiring one, with kicks to avoid and every hoof requiring close inspection. Here at Mt Creighton each of us examined about eight thousand hooves for the week.

On inspection of the hoof first you decide whether it has the "dreaded lurgy" by feeling the hoove's warmth or by its bad smell — or, if the sheep has it badly, the hoof will look rotten. The hard outer shell of the hoof is pruned back so air and a special formalin solution can sterilise it. If left unchecked, the hoof rots away completely; the animal can't walk and then faces a lingering death by starvation.

leave it, because its balance is so out that it will just fall over in the same position again. If it's that bad, you then put it out of its misery.

Another unpleasant task is foot-rotting — that is, pruning decaying flesh off a sheep's hooves. A virus gets in there and rots the foot to the core, and to fix it you have to prune away the outside shell. It's very smelly, and very painful for the sheep, so they kick mightily. But it has to be done right back: that is, the hoof has to be opened right up. We once had eight thousand sheep to look over. That's 32,000 hooves, an awful lot to do with just four people. We worked at this for about three weeks — a very tedious job. You roll the sheep from a bench into a dish so that they're upside down in front of you, and they're feeling vulnerable and kicking. It's the inside of the hoof that rots, not the shell. But it's the shell that has to be pruned away to let the air in and enable the foot to dry out.

SO much of mustering depends on events moving in a certain sequence, which in turn is achieved by experience and teamwork. If the sequence falters then the whole operation may be in jeopardy. This happened on a block I was working in the Wakatipu area. I was given the bottom beat because I'd done the top the day before. The bloke I was with, I'll call him Sid, was on the beat just above me. He was a young guy, just out of school; he had a few dogs and was doing casual work around the area. I was aware that he was inexperienced.

I stayed at the hut for about three-quarters of an hour while the others climbed to between 1700-1900 metres, and then I started. The man on the beat above Sid had hunted some sheep down and they had gone on to Sid's beat. That's the way it's supposed to work, except that Sid had already gone, so he couldn't collect those sheep. He should by rights have waited and made sure that the person above had gone through. So I spent about twenty minutes climbing up and getting these sheep, and I found a couple of other mobs that he'd missed. It was very rough country, full of bluffs, matagouri and manuka. I caught up with the others and then carried on driving my own mob of about 1500 sheep down below. My main job was to take the sheep in single file through the bluffs. I held a quiet dog on them, carried on and again came across Sid ahead of the other guys. He had again left sheep behind him and I stormed up to them. The country was a lot easier now and he shouldn't have missed those. I caught them, and he stopped then because he saw me climbing up on his beat. I put them across the creek and told him it wasn't good enough.

I was then to hold the front of the mob and take them down and keep them low so that they went through a wider track rather than spread out. So I caught the main mob and was proceeding to take them through when I was interrupted by Sid, who had caught some sheep but was not handling them properly. He was using his dogs too much, standing back when he should have been getting more involved. As a result, a simple situation got out of control and the sheep bolted back to where I'd just brought hundreds of my sheep around, and they were upset. They'd been split up and they were running back. I caught the sheep for Sid, then told him to bring them up. I told him that he must try and work within his dogs' abilities.

OF course, knowing what to do is one thing, but dogs are human too, so to speak, and when they're young their enthusiasm can cause problems. That was a problem for one of my dogs, called Hitch because he came from Hitchen Hills in North Canterbury. The young dog had never been run on the hill before. In paddock work he was excessively enthusiastic. I had to stop him whenever he got to a fence, because he'd attack the fence to get through it, especially if there were sheep on the other side of it that I wanted to move. Even if the fence was made of barbed wire it made no difference to Hitch. I could see the dog was crazy, so I had to stop him and persuade him either to go under the fence or to pause long enough to judge his jump to get over it.

When he went downhill he thought, with his lack of experience, that he could run as fast as he could on the flat, which of course he couldn't. I could see what was going to happen a mile away. When he started running I thought, my God he's going fast. He got faster and faster. He was trying to head off a bunch of sheep from going downhill. I was whistling like hell for him to stop; there was no need for him to try and head them off like that. But he had decided they were out of control — which is a bad fault in a sheepdog. Soon he was travelling downhill at nearly fifty kilometres an hour. Screaming through the tussocks, and the next minute he literally took off.

Hitch was flying. He did two somersaults. It was too much for me, I started laughing. Then he hit the ground and, like a cartoon character, he didn't flinch — he just launched himself off again. Wild! He did eventually stop the sheep, but what a way to go about it!

I once wanted him to bring some sheep down from a hill, so I sent him off up a bank a bulldozer had made, to reach behind them. Well, he shot up this bank and over the top, and there he was in mid-air. Then, rather than wait until he hit the ground, he started running in mid-air. It was priceless. Hitch wasn't talented at moving sheep, but his enthusiasm was fantastic.

Or there was the occasion at Coronet Peak Station when I was a little bit behind the mustering and working fast. Some sheep had just gone out of sight about a hundred metres away. I ran my heading dog out to the right, but the sheep didn't reappear. The dog didn't turn up, either. I thought it was a bit strange, so I ran another dog, one that was a bit more direct. I wasn't worried; I knew I'd catch them eventually. Next moment, these wild goats came over the ridge only yards from me and very nearly barged into me. And there was my first heading dog, straight-faced, bringing them in quietly!

WORKING with a team of musterers out in the backblocks and living in a hut was new to me when I first worked in the high country. One of the things I didn't realise was the amount of consideration that was needed. There would be seven of us in a small hut: six musterers and the cook, who was also the hut boss and who packed our gear to the next hut on packhorses. The hut might be only three metres by four.

The sleeping arrangements in those huts were such that a man had to catch forty winks wherever he could.

We all wanted "air time" to tell of our runs of the day and how our dogs had gone. Certainly we didn't miss the television or radio. The talk was nearly always constructive, in the sense that there wasn't much criticism of others. For my part, in those early days, I longed for one of the senior musterers to remark on my work, as they often did with their equals.

About six of us were mustering up Minaret Creek on West Wanaka Station. I had the second-to-top beat and we were stretched out in a line. The man on the bottom beat got into some trouble. Some sheep had been smothered through a poor bit of shepherding and we had a wait of four or five hours while the business was sorted out. We all dozed off a bit, but only lightly, because we knew how much work there was ahead of us. But our mate down below, Cooch, went solidly to sleep, and all the shouting from a distance couldn't rouse him. I could see his dogs running about and I thought he would wake up soon. But no. So I ran a dog down to him. In a way that was a betrayal of the dog's training; he trusts that when you send him after sheep that he can't see, they will nevertheless be there. But I sent this dog down to bark in Cooch's ear. He did too, and that woke him up all right!

Sometimes mustering becomes a waiting game while the others in the line catch up. Here my dogs Peg (left) and Bonny are having a quick snooze. When the muster gets going again, the other musterers bark their dogs up (that is, make them all bark loudly, as a signal). Sometimes I doze off myself, but my dogs still hear, and they wake me up.

It's first light as we head out for the muster on Glen Dene Station, above Lake Hawea. We're taking a rather round-about route to reach the comparatively inaccessible Wanaka Face.

Although a sheep only has to miss shearing time once to qualify as a woollie, some of them miss two or more shearings, and this tells you something about their character. The sheep being lifted at right has missed just one shearing, perhaps because it was weak, and so won't have developed any subversive tendency. Double fleecers are more tricky, as they are likely to have had some real practice in deliberately escaping the muster. But the ones below are triple fleecers, and really difficult; even though they may look docile, they're pretty close to being wild. Sheep which miss more than three musters are completely wild and just about impossible to bring in, but they're very rare.

One of the triple fleecers from the previous page finally makes it to the shearer. You can see how incredibly long the fleece is, though the wool isn't very good quality because the seasonal variation in food supply causes weak spots along the fibre. Because of this the wool is baled and sold separately. Above, the whole fleece spreads wide on the shed floor as Robert Dingle finishes the job, which has taken two or three times as long as usual because the wool's so thick.

WINTER

WINTER begins about halfway through May for me. The autumn muster is finished, a lot of the stock work is done, and I'm looking for another type of job — fencing maybe, or possuming. You're more relaxed; there's not the pressure of being part of a mustering team where each step you take affects the success (or otherwise) of the muster. You're feeding stock and looking for work that needs to be done; you're improvising. The attitude is more laid back. If it rains we might settle down for two or three hours, have a cup of tea and wait for the weather to clear, and if it doesn't we'll pull on parkas and go out and feed the stock. The necessity to get the job done is not as great; there aren't the same deadlines to meet.

In winter there might not be the work pressure, but there also isn't the company. In the summer you're working with maybe three or four people, but in winter people are taking off for other places: they go off to do fencing, to find contract work. So to that extent the seasons dictate your social relationships. But I can enjoy sitting in front of a fire, doing leather work and repairing my gear. I've made a bridle and a halter and I've repaired a few saddles. It gives me something to do.

Wethers gleefully make good their escape from a cold, high block on Temple Peak Station. Anyone who's ever had to wade through deep snow will understand why they're walking single file!
The first snow like this doesn't actually bother them much because they're in good condition, with plenty of fat to insulate them and provide food reserves. But if they are caught out for more than a few weeks they soon get pretty miserable.

At most of the places I've worked the boss' wife cooks for you in winter. You have your meals with the family, and then you go away and do your own thing. But even that varies a lot. In some places I'll stay on and watch television with the family, while on other stations television might be supplied in the quarters. I'd rather watch with company, although that doesn't happen too much. It's more common in summer when you're doing a job, there are shearers and the other musterers around. People might think the life of a high country shepherd is naturally solitary, but perhaps they don't realise the company that dogs and horses provide. They are a big substitute. You're hardly on your own when you've got half a dozen dogs around you. Okay, they might not speak English, but they're company nonetheless.

The first place I worked on, starting in February 1976, was Glenrock Station. I felt a terrific pride in doing a good job and in being a high country shepherd on the Rakaia.

I was there for two and a half years, and I learned a lot, from the Glenrock runholder, Charlie Ensor. His relatives run three places out there, and in fact the whole of one side of the Rakaia Gorge is farmed by two families. Having such a settled population makes the whole side of the gorge a close community. They didn't then have telephones so they had talking sessions by radio instead, five or six times a day. The closeness of the family engendered loyalty. You had the security and the pride of belonging to it — but you were also restricted by it in some other ways.

Dogs lined up for inspection at the annual Tinwald dog auctions, near Ashburton. This is one of several big dog sales held each year during the winter months.
It's not unusual to pay upwards of $750 for a good dog.

It's a Sunday, and Alan Hassel is cutting a slightly nervous Tony Stuart's hair with a pair of sheep shears. (Tony actually does want this haircut — he hasn't been pushed into it the way I once was!)

For example, I had longer hair than the others, and long hair was not one of the things you had if you were a high country shepherd. I didn't take it very seriously, but it was made clear to me that to be a good shepherd and part of the team, you had to have your hair cropped neatly. One night we were having a few beers and it became very obvious the guys were dying to cut my hair.

I'd said they weren't cutting *my* hair, and that only made them more determined to do it. I really didn't want it cut, either. I also resented being told my room had to be kept tidy. I thought, well what *can* you do your own way? The cook told us our rooms would be inspected, but there was never anything done, so where that idea came from I'm not sure. Anyway, this night I was standing against the door. I could feel the atmosphere. I felt that any minute an attempt would be made to cut my hair, and I wanted to be able to get out quickly.

Then after two or three hours standing by the door — I was armed with a broom handle too — I was nicely tricked into turning on the heater. As soon as I turned my back there were two guys in mid-air. There was a little fight. They started to cut my hair with shearing blades. I lay still for a while then made a burst, pushed a couple of guys off and barged out the door, damaging it in the process. So I ended up outside with my hair half cut, and a sprained ankle. I had also inflicted two black eyes. At the breakfast table next morning there was some embarrassment, and quite a few disapproving looks.

I suppose that sort of thing still goes on, and in fact a year later I participated in a very similar thing, holding a guy down and cutting his hair; and he wasn't even a shepherd, he was a tractor driver.

I never grew my hair long again. I heard of a guy, I never met him, who did grow his hair long again, and feeling against him ran quite high among some of the shepherds. He just wanted to grow his hair long, and he was prepared to pay the price of being a bit lonely because of it.

It might be easy to sit in a city and say, crikey, look at those guys, what a bunch of barbarians they are, cutting each other's hair like that. But at least we didn't colour our hair purple and set it in spikes.

The following day I was almost back to my old self. I looked a bit of a drongo, with half my hair going down past my collar in ragged bits and short on the top; but I wasn't terribly worried. I don't know that there would have been any such initiation if I'd gone there with short hair. I think I'd have been watched, and it would have taken a while to thaw out and see if I fitted in. I was annoying my workmates by wanting to belong to their group yet saying I was going to be different. So I brought it on myself. Later I participated in scragging other people. We were sticking up for what we stood for rather than having newcomers believe they could be different and still belong to the group. They couldn't. You were a musterer and that was that. We heard on the grapevine that the bosses weren't too happy about initiation rites, but my guess is that the protests might have come more from the wives than the bosses. They probably thought it had gone too far.

ONE of winter's toughest jobs is finding and freeing snow-bound sheep. In spring, the grass starts growing at the lower altitudes. Throughout summer the growth creeps up the hills and the sheep follow it. By the end of summer they're used to the fresh growth being on the tops, and some sheep go back up after they've been mustered and crutched, in April, if there aren't any fences, looking for that grass. It takes them only two or three hours to walk up into the snow range, and what happens is that if a heavy snowfall comes, say two or three feet deep, the sheep get trapped. The snow's too deep for them to wade through. Wool really demonstrates its qualities here. I've seen sheep imprisoned in

This dog's just been bought at the Tinwald dog auctions and is hanging back a bit because he's not too sure about his new owner. In the background a dog is being auctioned while the owner demonstrates its capabilities at the same time. Depending on how the owner and dog get along, they may learn to work together efficiently in no time at all, especially if they do a lot of work straight away. Or it may take them a few weeks to get used to each other. If the dog isn't working well after a month it'll never work for that particular shepherd and the only solution is to sell it again.

Right: Mt Cook dwarfed by a nor'west sky with hogsback clouds building up — a sign of bad weather on the way. The shepherd always keeps an eye on the sky as well as listening to weather forecasts, and combines both sources of information to get a clear idea of what's in store. When you look out to see a sky like this it usually means a complete turnabout of whatever you had planned for tomorrow.

Far right: A red deer stag on Mt Creighton Station, with the snow-capped Humboldt Mountains in the background. I enjoy mustering deer because they're more sensitive than sheep, and so present a challenge. They scare more easily and move sooner than other stock, even when they've been a long time in captivity. This stag is unusual in having fully formed antlers during the winter — usually the antlers are cropped at the velvet stage, during summer, because the velvet is extremely valuable.

these conditions, with hardly anything to eat, for six weeks; I've been told they will actually eat each other's wool. On one occasion a mob was found still alive after two months of being trapped. Apparently the snow was so deep it covered them, keeping them insulated, and that's when they ate each other's wool. It was a sad irony that, after they had been uncovered, they died. Perhaps their weakness and exposure was too much, and they were unable to survive the transition to a more normal environment.

At Glencoe Station near Arrowtown I was once involved in rescuing sheep which had been trapped in snow for six weeks. They were imprisoned in a thirty-square-metre area. All the snow tussock had been eaten right down to the crown of the plants. The sheep walked out once we had made a track for them. They were Merinos, which are probably the hardiest sheep breed, though they're by no means the most aggressive.

Breeds can be so different that they hardly all qualify as sheep. One of the biggest differences I've found is between Merinos and Perendales. Perendales will stand there and look at you, and wait for you to make the next move. And you *have* to make the next move, and it has to be a sound one if you're to get those sheep in. Merinos are not as defiant. They'll move away if you bark up, and they'll keep going all day, whereas Perendales will stand and watch you. You can bark your dogs up, and so what? You're way down there, they think, you can't do anything to us. They'll test the situation out, whereas Merinos will move away.

The differences are accentuated by the way sheep are handled when they're young. You go on to a station where the owner believes you shouldn't use a lot of hunter voice and consequently when you do bark up you get a great effect. The ability of the dogs is important, too. If the farmer's attitude is that the sheep should never beat his dogs, the sheep will develop respect for them. Exactly the opposite happens if a farmer is quite sloppy in his

attitude towards getting every last sheep in. The sheep soon learn that a little bit of effort will leave them free.

Anyway, on this particular occasion, at Glencoe, we set out to trample a track through the snow for the sheep to follow once we'd freed them from their snow pen. A good track is one which the sheep can follow naturally — that is, it will lead them around hills, not straight up or straight down. This more natural track helps them make the break from the predicament which they've come to accept, being naturally quite passive animals. With us there they're on edge anyway, and if the track is persuasive and if we're quiet, there's a good chance they won't actually panic and flounder out in all directions, causing us an impossible amount of work. Sometimes you've got to pick them up physically and carry them. If it's uphill it can be hard work through two or three feet of snow. They're getting weaker all the time of course, and they can't stand much fighting through snow themselves. They'll try and find a way out, but they are just not up to it.

The track we made was really only about a hundred metres around from the sunny side of a ridge. I stayed at the bottom of the snow pen while Brian, the boss, brought them around to the track we'd made about halfway up the side of the pen. That way I was able to hold the bottom and manoeuvre the sheep so they'd see the track. We held them in the pen, in close proximity to the track, for some time so they could settle down and get used to our presence. This patience paid off, because eventually they just trickled off, and because of their calm state the last ones didn't panic as they sometimes do. We kept very quiet, because it was their escape and they were having to turn their backs on us. To move around too much while they were walking away would cause them to rush — to almost certain disaster. So it was a very delicate situation.

On another occasion there was a mob that had been trapped in there for six weeks, and I had to tramp a track only a short

These sheep had good reason to be looking towards the camera for advice as to what they should do next, because I had just freed them from a month's imprisonment in deep snow. They had missed the autumn muster and stayed up high through a succession of snowfalls that piled up and became too thick to walk through until they froze solid. At this stage the sheep could have walked over the top of the snow, but they stupidly didn't think of trying — instead I had to show them the way out.

Canada geese near Mt Cook. These introduced birds have become a feature of the high country and although they're beautiful, they're also a pest. They always go for the grass in the best paddocks, or young pasture which is still being established, and their very acidic droppings foul the pasture.

The only practical way of getting sheep down when they're trapped in soft snow is to trample the snow down hard with your own feet. The technique is called snow raking. Here the sheep are being eased on to the track. This is a delicate operation because if they panic and flounder off then you have to start all over again.

*We'd walked most of the way up this hill in the Crown Range to help some
sheep out of the snow and I thought it'd be a pity not to go
all the way up to catch the view.*

Bill Dagg (left) and Guy Bellerby are two shepherds I've worked with a lot. The dog's name is Guy, too — needless to say he's not one of Guy's dogs. (If that sounds confusing, then imagine how difficult it must be for the dog when the two men are talking to each other!)

The kea is a bird I love and hate. When it fearlessly comes up to me and hovers over my head in a howling gale, displaying the brilliant orange colour under its wings, I can only love it. But I hate it when I see the cruel holes it pecks in the backs of live sheep.

One of the reasons why keas are so controversial is that there's no photographic evidence to support the belief that they attack sheep. This is because they do it at night. The one time I've actually seen a kea on a sheep's back was at night, in the car headlights.

Heavy snow on Glencoe Station during July. It's quite high here and the snow's about a metre deep — no place for man, dog or stock.

Cotton plant flowers coated with ice. There is probably nowhere else in New Zealand where temperatures fluctuate between such great extremes with the seasons.

Glencoe Station doesn't keep breeding stock because the severe winter and spring there make it
too difficult to breed good stock and grow them to a decent size. These hoggets have been
brought in from another station, for wool production, and until the snow clears the only food
they get will be what we feed out to them — hay and oats.

These three competitors in the Dog Derby have finished sending their dogs through the first markers and are now on the section where you race down the ski field with your dog.

The starting line of the annual Queenstown Dog Derby, held at Coronet Peak ski field. Competitors line up behind that barrier, then try and send their dogs up through a set of markers. It's very difficult because everyone is whistling and shouting at the same time and the dogs are replying with barks of great excitement.

This is Sid, one of Bill Dagg's huntaways, who is a hero among the other dogs at his home on
Coronet Peak Station because of his strength, his ability to work hard, and his sheer gutsiness.

distance, maybe fifty metres, but it was uphill, and damned hard work. And the frustrating thing was that all this work might be wasted. This mob was on the bottom of a ridge and I had to get them to cross a frozen creek, where the snow was a metre deep. There was no way they could get through that creek: they'd just be cast, they'd have no traction at all. So I had to tramp a track into the creek and out the other side quite steeply. I travelled about two metres in ten to fifteen minutes, my dogs in behind wondering what I was doing. It took me three to four hours to trample the track. And when I got close to the sheep, how would I convince them that I was rescuing them — not threatening them? I knew from experience that those three hours of hard work might be completely wasted in a matter of seconds by a bad move from a dog, and that put me on edge. I stayed back and let the sheep become aware that I was there, letting them see me. When I was convinced they were quiet enough I proceeded to take them to the track. This time they led out and I just stayed back where I was. When they filed out the other side they gave, to my amazement, little bucks of joy, tossing up their heels. I didn't think they'd have it in them!

FROM one extreme to the other, snow to fire. It was in the late winter of 1983 that a fire got away on us. The runholder, Aaron Radford and myself were contemplating burning a block of fern, with the idea of putting super and seed on afterwards and then fencing it. The question was, should we? There was a slight breeze blowing, but to counter that, the tops had a good covering of snow to protect the fragile grass up there, and the fern was dry. We got the necessary permission from the Forestry Department, so we went ahead and burned. Really, the minute we lit the fire we knew it was going to be a good burn. The fern just took off. It hadn't had a burn for a long time, and it was very dry. The fire roared up the hill.

It burned throughout the day, which was expected, and later, in the evening, I stayed behind for an hour and went and checked the fire from the road. It looked to be quiet. It had burned the whole hillside and was just simmering around the tops, where there was snow.

So everything looked under control. However, later on in the night Aaron's wife reported that the fire was still burning up the top. Aaron felt a bit uneasy then. He went up to check it at about four o'clock in the morning and at that stage it still appeared to be burning right on top, about a thousand metres up.

In the morning I was woken up with the news that we were off fire-fighting. The fire had got away. What had happened was that the fire had gone up the hill, crept along a small ridge, and then gone down into another valley. We were lifted into the area at about nine o'clock after the Forestry men had been lifted in with their pumps. That was frightening. The wind was rough, and the helicopter was bouncing up and down three to six metres at a time. We must have resembled a sandfly, veering around as we were. The pilot did a remarkable job though, landing with one skid on the side of the hill.

The question now was, where do we go? We were looking at a burning front of about three hundred metres, from the top of the valley to the bottom where the Forestry men were. We were thinking about what was best to be done, when the wind suddenly picked up, and the fire just exploded. We both went white. If we had decided to go down to fight the fire we'd almost certainly have been burned to death.

Within half an hour the whole picture had changed completely. It was as if someone had turned the oxygen on. There was nothing we could do, and we were lifted out. It was eventually put out with the help of farming neighbours, by back-burning and by about six helicopters using monsoon buckets.

Aaron Radford and I lit this unwanted fern to clear the ground for planting in pasture, but we ended up with more than we bargained for, because we misjudged the strength and direction of the wind. Also, we didn't think the fire would burn across the ridgetop because the vegetation on top wasn't the sort you'd expect to burn. But the fire ran out of control and burned a neighbouring valley. All the land in these pictures was burned in just a couple of minutes when the wind suddenly, unpredictably picked up. It was absolutely awesome.

IN April of 1977 two Japanese photographers came around when I was working at Glenrock, up the Rakaia. They said they were doing an advertisement and they asked us to model for them. I was with three other musterers at the time. We were having a quiet Sunday, a few beers, and thought, well okay, there's no harm to it. So they took our photographs and went back to Japan to assess them and we didn't think any more about it. Three weeks later I got a phone call asking me if I wanted the job. It was a bit of a shock, but I accepted. They came out again, all of them this time. There was a camera crew, stills photographer, script writer and producer. They wanted a big mob of sheep rounded up with a fierce nor'west sky as a backdrop. We collected about a thousand sheep and I found it hard telling the director that I couldn't bring the sheep right up to me, I couldn't pat them. They were quite frustrated that they couldn't get the sheep close to the camera and me close to the sheep. I presumed that they had got that idea from other shepherds around the world who are with their sheep all the time. I tried to explain that these sheep had just come off the mountains and hadn't seen people for six months.

I was given a walkie-talkie in my hill bag, I had my old swannie on, and I looked like a real rough diamond. I was to take these sheep past the camera in a nice flowing motion, and walk close in behind. In fact they told me to run in behind, so I was close, much against my better judgement, because you just don't push sheep when there is no need. My dogs thought I was going crazy and they got a bit excited because it was unusual behaviour. I had to go round in a half-bend for the film crew, and it was difficult to get a thousand sheep to move the way the crew wanted them to.

This went on half a dozen times, with me getting quite frustrated, calling and whistling to the dogs to make sure they did exactly what was wanted. The crew was delighted with the performance. At one stage they wanted me to swear at the camera but I didn't think that was a very good idea. Then everything started going right. I was bringing the sheep on and I could see that they were well in control; the interpreter was encouraging me through the walkie-talkie and everything was going really well; "Just keep it up there, in behind, keep close to the sheep, that's very good." I sent a dog to curl the sheep round to the left. Peg was her name. I knew I was pushing the dogs hard: they weren't used to this continual repetition, they were getting sick of the job, and because the pressure was on them I knew I was pushing them to the limit.

Then it all broke down. Right in front of the camera Peg squatted down and relieved herself, much to the consternation of the film crew. I thought it was all very funny, until I realised I would have to do it all again. It was a commercial for whisky, and I never did get to see it, though they did send me some posters.

IN the Queenstown area they have a winter festival which tries to involve all the different groups in the area. One of the events is a dog derby. In this, farmers compete with their dogs on the skifield for quite a good prize — three or four days in the North Island. Well, I guess it's a good prize!

I thought the festival was a great opportunity to meet the people in the area. I was pretty familiar with skifields anyway, and the idea of going up there with a dog and acting the fool sounded good. You're given a number to wear and then you get on the ski-lift. Your dog has to walk up, which sounds simple enough — except that there are a whole lot of skiers around and these dogs have never seen skiers in their life before — people screaming around on bits of wood, it is quite daunting for the dogs. In addition the lifts are going round, coming and going, the colours are kaleidoscopic and everybody seems to be shouting. The dogs are thoroughly bewildered.

We were taken halfway up the ski-lift and assembled on a flat piece of ground with a steep hill in front of us. There were farmers

These pictures were taken on a cold, snowy day when Guy Bellerby and I were taking some wethers out to a winter block on Mt Creighton Station. It snowed all day, without once stopping.

Right: The modern way of feeding out during the winter. Here at Glenrock a large automatic trailer dispenses haylage (grass stored semi-dry and handled only by machinery). The trailer is shared among three stations and enables one person to feed all the sheep in one morning, whereas the job used to take six people in the days of small dry hay bales, which had to be manhandled individually. Haylage is better, more succulent tucker, too.

and shepherds, and farm dogs and town dogs all running around and having a great time. The idea was to send your dog up a hundred and fifty metres, through two poles then back to you again. That's simple enough in an everyday farm situation except that here, thirty-five dogs and people have to accomplish that at the same time, through the same markers.

I was keyed up for the start, but when it came the chaos of commands and dogs racing on snow was just too much — I cracked up laughing. I kept whistling the dog up and I think it was more fluke than anything that she went up. For her it must have been like being on another planet. When a dog is told to go up a hill to collect sheep it is usually on its own, but this time there were thirty-five other dogs around. It would be debatable whether the dog could even hear your whistle. I know they can hear your whistle in among, say, a dozen other whistles. But among thirty-five different whistles and yells?

Once your dog comes back to you, you have to run a kilometre down the ski-lift line as fast as you can. I had already been given a tip to put on waterproof trousers, so you run as fast as you can then sit on your backside, getting your breath back as you slide. I was running pretty fast, and when I sat down I went faster and faster until I was out of control. I had a walking stick with me which I tried to dig in, but this didn't have much effect because the snow was packed down for the skiers. The dogs kept up as best they could, thinking the whole thing was crazy — which it was — barking all the time.

You race down towards the terminal at Coronet Peak, with a lot of people cheering and yelling, and then when you cross the line,

you have to stand and send your dog again up the same distance, through the markers and back through them, just the same as the first time. The dogs hear the whistle and go, still thinking that this is barmy because there are no sheep up there, but they do it anyway. Then you call them back to you again and from there you rush down through the flags. I got second and won $150 worth of liquor. That year the dog didn't get anything except a big hug, but the next year a butcher gave half a side of hogget.

There is another contest, one I didn't compete in, called the dog barking competition, which is held in one of the local hotels. It also is bedlam. They use the band's public address system for the organiser to be heard. He gets the competing dog and its owner inside in front of the crowd, and then they're given the floor. The dog is to make as much noise as possible, though I think that some of the performances of owners trying to get their dogs to bark should earn a mark or two. Some hiss while jerking their heads in mock pursuit of a cat and others wave their arms around in a way vaguely reminiscent of an insane orchestra conductor. Others clap, while one or two tease their dogs by pulling their ears.

One person there had a backing dog, which is one that's been trained to run over the backs of sheep and frighten them up a race into dips and into other areas they'd prefer to avoid. Then the dog drops in among the sheep and walks back, and that packs the sheep up. Anyway, at this dog barking competition the man gave his dog the command to go back, and the dog jumped on to a table then on to people's shoulders. He was so quick, before we knew it he had gone three or four deep into the crowd!

They're off in the Queenstown Dog Derby! The start is the worst part because there are so many distractions all around, like colourful skiers.

Before the Dog Derby the dogs have to run up the ski field while their masters take the easy option and ride the chair lift. The dogs take a while to catch on to the idea of receiving their commands from up above!